Process Development Division 2018

Core Programming Area at the 2018 AIChE Spring Meeting and 14th Global Congress on Process Safety

Orlando, Florida, USA
22 - 25 April 2018

ISBN: 978-1-5108-6446-7

Printed from e-media with permission by:

Curran Associates, Inc.
57 Morehouse Lane
Red Hook, NY 12571

Some format issues inherent in the e-media version may also appear in this print version.

Printed by Curran Associates, Inc. (2018)

AIChE
120 Wall Street, FL 23
New York, NY 10005-4020

Phone: (800) 242-4363
Fax: (203) 775-5177

www.aiche.org

Additional copies of this publication are available from:

Curran Associates, Inc.
57 Morehouse Lane
Red Hook, NY 12571 USA
Phone: 845-758-0400
Fax: 845-758-2633
Email: curran@proceedings.com
Web: www.proceedings.com

TABLE OF CONTENTS

Some abstracts in this book refer to images that were not made available by the publisher

514424 Best Practices for Accelerating Process Development with External R&D Partners

Abstract:

Chemical process development continues to deliver value across the spectrum of chemistry-related industries, including petrochemicals, bio-based chemicals, specialty chemicals, energy, and pharmaceuticals. Process development provides a focus for reducing costs, both CAPEX and OPEX, and improving product quality and process robustness. However, the investment to have the necessary capabilities in research and development can be significant. This is a challenge for start-up companies who are raising capital to develop their technologies; it is also a challenge for established corporations trying to maximize the efficiency of their work force. Recently, more companies address these challenges through external research and development programs, such as joint programs with research universities, other companies, or through contract research and development organizations.

External research programs allow companies developing new technology to leverage their internal resources with additional expertise and infrastructure that is already in place at the partner's laboratories. Process development can require deep knowledge in various fields, for example corrosion, crystallization, and new cutting-edge technologies, and requires the experience and knowledge of the challenges that are typical of new technology implementation. Maintaining these skills, even in large organizations but certainly in small start-ups, may not be cost efficient or possible. These skills can be contracted on an individual basis, but this often does not provide the deep understanding of the process that a dedicated technology development team provides. In addition to the expertise, the infrastructure for the facilities requires nearly as much investment as the technology development. This investment in infrastructure, although critical to the technology development, does not generate additional value to the technology and will not be recoverable after the technology has been implemented.

However, contracting outside one's organization for the technology development can leave a company open for value loss. Technology value can be lost through the loss of intellectual property ownership – either due to inadvertent exposure outside the parent organization or contractually through joint IP ownership between the technology owner and the external organization. Other opportunities for value loss include losing important aspects of the technology as the development is handed-off between different external organizations as the technology progresses through the various stages, the so called "white spaces".

This presentation will discuss the challenges of process development and how the various external research organizations can improve the efficient usage of investment dollars in new technology commercialization.

The Need for Pilot Plant Specific Standards

I appreciate that the Process Development Division has again asked me to give a Plenary talk on Pilot Plants. While I felt that a talk on Sex in Pilot Plants would have vast appeal and wide interest, that idea failed when I was unable to come up with anything to say after the title line. So I am afraid I am going to discuss a less potentially salacious topic but nevertheless one that I think is of great importance to our work.

So, let me return to a topic I first talked about in 1993: Pilot Plant Specific Safety Standards.

Trying to design anything and ensure it is always safe, is a difficult task. While designing the unit, you need to identify all the potential hazards, access if they are a credible risk, and provide appropriate mitigative measures for those that are. Every mitigative measure you provide adds to the cost and schedule so you are continually trying to ensure that you are protecting against credible scenarios, ones that have a realistic chance of occurring, versus incredible or highly unlikely scenarios. And during the entire process you are always worrying that you may have missed something or failed to address it adequately.

Commercial plants have numerous codes, standards, and recommended practices to help them ensure safe operations. (For convenience I will call all of these "standards" during this talk.) They are available from numerous organization such as ASME, NFPA, API, CGA, ANSI, ISA, and many others. Following these standards helps the designer produce a safe product. The group that developed the standard thought through the potential hazards and developed approaches that, they feel, should mitigate the attendant risk. The designer does not have to spend time and effort analyzing many areas; a proscribed safer course is usually given in some detail. Confirming they "met the standard" is usually sufficient.

However, virtually none of the standards developed by these organizations are directly applicable to pilot plants. This leads to issues in the real world and tremendous inconsistency as each pilot plant designer tries to analyze their unit and develop their own answers.

A pilot plant designer will follow NFPA 70 The National Electric Code because it is a legal requirement and, in most cases, can reasonably be interpreted as to how it should be applied to a pilot plant. There will be issues generic to pilot plants, that may need to be addressed (specialty equipment, mechanical protection for all wiring, specialty assemblies, etc.) that may require some thought, analysis, and determination of how best to comply but – generally – it works.

Most organizations will apply the ASME Boiler and Pressure Vessel Code to vessels used in pilot plants, often even to vessels less than 6 inch ID or to those fabricated in house or by non-ASME approved fabricators. But here the road to compliance is more uncertain. Is a "vessel" made of 1 inch or 2 inch standard piping components a vessel or just a piece of pipe? How about if it is made partially or wholly out of compression fittings? What if some of these components are welded to avoid leaks? What about wanting to use non-ASME approved components that are important in research whether vacuum flanges, compression fittings, or similar items? In these cases, applying the ASME code becomes more challenging. It is no longer just a matter of reading and following the standard. Now some analysis and determination of the best approach to comply is required. And different designers in different organizations at different times will take different approaches. Not quite so "standard".

Does ASME B31.3 Process Piping, which recommends a pressure test of all new piping, really address the concerns of non-welded compression fittings? Is it significantly safer to require a higher than operating pressure hydro or pneumatic test that requires removal and replacement of significant components (relief devices, gauges, and possibly some other equipment)? Tests I participated in in the 1980's clearly showed that compression fittings would leak horribly and noticeably under a leak test long before there were any chance of them failing even when improperly installed. Here, the components – compression fittings – and the normal pilot plant practice – rigorous leak testing at 80% or so of operating pressure – are not really addressed by a process plant code. The best answer is much less apparent and different organizations will decide on different approaches.

NFPA 45 Fire Protection for Laboratories Using Chemicals and AIHI Z.9 Laboratory Ventilation provides excellent guidance for laboratories but what about laboratories with significant numbers of pilot plants in them, either in hoods or larger standalone pilot plants out in the laboratory? NFPA 45 is quite clear that is does not address pilot plants; Z.9 never mentions them. NFPA 45 and local building and fire codes clearly establish overall limits of flammable materials inside a laboratory space but does the amount of flammables outside a hood in a standalone pilot plant suggest a different approach? Both NFPA 45 and local fire codes don't address the issues with the higher pressures and temperatures typical in pilot plants. Should this suggest more stringent limitations or additional mitigative measures? Most research organizations will adopt some limits but they vary enormously and their rationale ranges from fairly through to blatantly casual. Lacking a consensus standard to fall back on leaves everyone to their own devices, often with unfortunate results.

NFPA 497 Recommended Practice for the Classification of Flammable Liquids, Gases, or Vapors and of Hazardous (Classified) Locations for Electrical Installations in Chemical Process Areas gives numerous examples of how to electrically classify an area. The distances are based on installation sizes where SMALL or LOW is defined as less than 5,000 gallons in size, 100 GPM in flow, and 100 psig in pressure. The first two values are usually enormously larger than all but demonstration units while the last is often far exceeded in pilot plants. How relevant are the distances suggested in these cases?

ASHRAE 110 Methods of Testing Performance of Laboratory Fume Hoods gives excellent methods for testing laboratory hoods but no real advice on how to use them safely, what should or should not go in them, or how to deal with common problems. They also do not give pass/fail criteria for their tests except in the most broad and unspecific way. Not very useful to a pilot plant designer with a problem hood.

Other codes such as NFPA 30 The Flammable and Combustible Liquids Code, or NFPA 58 Liquefied Petroleum Gas Code give solid guidance on storage and handling of flammable and combustible materials or LPG's but never mention pilot plants or laboratories. And many of their guidelines are irrelevant or at least incredibly difficult on a pilot plant scale. Does one just ignore them? Does one try and pick and choose appropriate sections? Does one try and develop equivalent quantities, approaches, or guidelines? Does one just note that they "don't apply to pilot plants or laboratories"?

NFPA 55 Compressed Gases and Cryogenic Fluids Code gives excellent guidance on the storage and handling of hazardous gases and cryogenic fluids. Yet it only mentions laboratories to defer to NFPA 45 for lecture bottles and educational and instructional laboratories. Is there a difference as to how many cylinders can be stored in a laboratory versus a different space? Is the full fire separation really warranted between a single cylinder of a flammable and a single cylinder toxic material in a well ventilated laboratory? NFPA 2 The Hydrogen Technologies Code clearly intends their standard to cover use of hydrogen "in all occupancies and on all premises" but does not address laboratory or pilot plant applications. Both standards are widely used and provide excellent information but effectively only address plant and commercial applications.

Even a cursory reading of many introductions and histories of these standards shows that the purpose for their development was plant scale applications. Their membership rosters demonstrate that this remains their focus. The areas they cover are those applicable to and of interest to plant scale processes. Attempts to get them to add sections on pilot plants usually fail as they are of no interest to the groups developing them or the groups acknowledge they do not have the skill and experience base to address these issues.

So what do we in the pilot plant world do? We often read the codes and try to apply them to our work. This may be as easy as deciding even a 3 inch ID vessel should be fabricated to ASME standards or even deciding to use an ASME fabricator to try and ensure higher quality. Of course, the fabrication requires that you find an ASME fabricator who is willing to make this small size vessel. It also requires that you forgo the ability to ever significantly modify the vessel in house. These, and similar problems often result in a decision to design the vessel to the standard but use a non-ASME fabricator or in-house fabrication. This, of course, raises the potential that not all other aspects of the code – which perhaps should be of concern - may be fully followed.

How about trying to electrically classify an area? Does a 50 gallon tank of flammable liquid require the same distances as recommended for a 5,000 gallon vessel? Should one reduce them to 1% based on volume? What if the tank is operating at 2,000 psig or four times the value for high pressure? Should one increase the distance by four? Or two? Or assume it balances out the smaller volume? Should I even bother to classify the area around a 50 gallon tank? A 5 gallon tank? A 50 cc container? Clearly at some point NFPA 497 or API 500 recommendations simply make no sense.

What about those areas that constantly arise but which don't appear to be addressed anywhere? How much flammable and/or toxic materials can I have in a pilot plant before I should consider classifying the area or providing special ventilation? What sort of ventilation should I have for an area with pilot plants handling only flammable materials? Or handling toxic materials as well? How much does a gas detection system mitigate hazards?

These and numerous other questions plague pilot plant designers and safety professionals every day. We are forced to come up with our own answers in splendid isolation. At best, we may be able to call a colleague or two, attend a course, read an article, or hear a presentation about what another organization is doing. Information is limited, the rationale for the decisions often unknown, and little open discussion occurs. No real attempt is made to try and reconcile one organization's decision to allow X versus another's decision to allow Y. When pressed most simply must fall back on "that's how we do it here." Less than a very unified approach. And if, heaven forbid, anything goes wrong even if no one is hurt, the organization or designer is left trying to explain a decision or approach that, at that point, everyone now overwhelmingly recognizes was wrong or at least inadequate.

So, what do most pilot plant designers and operators do? Sometimes we try to follow the letter of the code despite the obvious problems. However much we try, my experience suggests we almost always must compromise on some requirements that simply can't be met effectively on a smaller scale. Alternately some designers may review the code and select the portions that they feel are most relevant and should be followed, modifying or ignoring the rest. This approach entails a potentially higher risk. The sections modified or ignored may be addressing an issue that thus does not get mitigated. The rationale for the selection of sections to be followed may be weak, inappropriate or based on an incomplete understanding of all the potential hazards. Any selection process always has to assume some sort of basis; if this basis is not always true later than the selection process – however well performed initially -may produce a higher hazard down the road if the assumed basis is no longer the case. Deciding to reduce the recommended distance to be electrically classified to a 5 foot radius around a 50 gallon drum from the 15 or 25 feet of a SMALL (i.e. less than 5,000 gallon) plant installation may well make sense. But what happens when no one relooks at the installation years later when a 200 gallon IBC has been substituted as it reduces the hazards associated with handling a lot more drums? What happens if the 5 foot radius seems reasonable because you know that actually none of your tanks currently in service are over 10 gallons and you picked the 55 gallon figure as a boundary without any real analysis of what that larger quantity could produce? And the basis for the selection and analysis is usually poorly if ever documented and usually lost in the organizations records right after promulgation. Over time, organizations tend to forget the basis and or the selection process, or misinterpret them, or slowly but significantly modify them and assume they are still following the complete standard. This can create credibility issues when an incident or an audit points out the reality. Alternately the organization feels they "substantially" comply with the standard, or comply with the "principles of the standard". These nebulous declarations do not offer much ability to determine how effective or realistic is the actual compliance.

Reading and understanding many standards is often less than a straightforward task. Definitions used in standards often vary between different standard making organizations. How the standard is written, organized, and how easy it is to understand varies similarly. Standard language tries hard to be through, concise, and enforceable; sadly, the result often borders on trying to decipher a foreign language with only a pocket dictionary. Many standards tacitly assume a given basis or common approach. While this is obvious to all the standards committee members, it is less obvious (if obvious at all) to someone without their experience. Standards language, however carefully developed and thoroughly reviewed, is often open to alternate interpretations. Does the exemption to section 1.1.3(b) include the table as well or only the specific written requirement? Does 3 feet meet the requirement for it to be "immediately adjacent"? Interpreting a dead language often is not much harder. The time and effort required to gain adequate familiarity with the code is also a problem. Many research personnel simply cannot devote the effort to fully understanding the code to evaluate its basis, concerns, approaches, and application effectively.

Some organizations turn to developing their own internal standards to bridge the gap. These may be casual, somewhat arbitrary guidelines, hard rules based on a specific calculation, even undocumented standards that are nevertheless followed because everyone "knows" the accepted limitations or any combination of the above. These internal standards may be applied to an entire organization, a part of an organization, a specific site, or a particular group. Other sites or other groups may follow very different standards. The basis for theses standard is often unclear. The logic used to analyze, and risk assess their contents less rigorous to almost non-existent. Many have, in fact, grown up by picking the highest current value. "All hood units are limited to 5 gallons" sounds less reasoned when one finds that the largest hood unit is only 1 gallon. "No use of those compression fittings over 1 inch" sounds less relevant when the largest ever used on site is ½ inch. Yet, to be fair, many commercial standards grew from the same starting point. But lacking a wider base for consensus there is a much higher risk of these internal standards being overly conservative or requiring inadequate mitigative measures. These standards often become enshrined as holy writ over time and require significant effort to update or revise for current needs, often because determining the original basis or analysis is very difficult to impossible to determine. In numerous cases, I have found that the basis applied to a concern or a situation that no longer exists, yet the restrictions or mitigations are still religiously followed.

When I worked for ExxonMobil Research and Engineering we had a large and fairly comprehensive set of pilot plant specific standards which had been developed and maintained over almost 50 years. As a large company with a significant research effort, we believed this program was worth the cost and effort. Nevertheless, at various times, the work was essentially shelved due to tight budgets or limited resources. The comprehensiveness of the entire body of standards was limited. We never were able to develop standards in all the areas we really needed. Limited time, funding, or resources, lack of common agreement on the content, widely varying operations, and different unreconciled approaches all contributed to making it impossible to develop some standards even when we all felt that they were needed to address a problem area. Many of the standards were poorly documented as to their rationale. All too often (even if that rationale was simply "we don't need more than this right now"). And the entire effort was always subject to inherent problems ranging from criticism over their content (too strict, too lenient, too complex, too simple, etc.) to being ignored because these pilot plant specific standards did not follow companywide standards for large scale commercial plants. (Of course, how one puts a ½ inch schedule 80 thermowell on a ½ inch OD reactor was not explained. And an honest discussion of what such a heavy assembly really contributed to the safety of a research operation often got lost in the politics of criticizing the company's approved commercial scale requirements.)

Developing even guidelines – never mind standards - for pilot plant operations is difficult and time consuming. AIChE's Center for Chemical Process Safety has been developing a book on GUIDELINES FOR RISK BASED PROCESS SAFETY IN LABORATORIES AND PILOT PLANTS since 2013 and it still looks like it will be a year or more until completed. Rich in case studies and example approaches, even when completed it will be more of a reference identifying areas and concerns that one should evaluate rather than giving any specific guidance. It will leave the onus of deciding exactly what to do in your specific application firmly on the designer's plate.

So, having raised this issue what do I suggest (aside from copious quantities of alcohol consumption after work)?

I think that AIChE and the Process Development Division should take up the call to try and develop a set of consensus standards for pilot plants. A group of interested and knowledgeable practitioners should be gathered and asked to develop a structure for these standards and defining the major areas they should cover. A prioritization within this structure would probably also be prudent. Then individuals or groups could volunteer to begin drafting a proposed standard which is then reviewed and "blessed" by the group. This could, within a few years, result in a significant number of recommended practices for everyone to use. We would have to resolve how these practices are disseminated and made available for use. A book? Pamphlets? Web based? All of the above? We would need to decide how they would be maintained and updated. We would need to address how they would be best applied in different industries. These and numerous other questions would have to be discussed, alternate views aired, alternate approaches posed and discussed, and a consensus agreement reached. Hard work? Yes. Tedious and time consuming? Yes. But incredibly valuable at the end.

I encourage all of you listening to consider supporting an effort to ask AIChE to take up the lead in this area. This is a real opportunity for the Institute to make a true impact in a way useful to many of its members. I am ready to volunteer; I hope this talk has convinced you to do so as well. Thank you.

514441 Innovation in Packaging to Address Market Needs

Abstract:

Protective properties of plastic make it one of the worldâ€™s most sustainable performers in delivering environmental, economic and social value. Market drivers such as demographic and lifestyle changes, improved consumer convenience, enhanced shelf appeal, increased shelf life, reduced food waste, and sustainability are putting increasing demand on materials suppliers, converters and brand owners to continue to deliver innovative packages. These market drivers and associated performance requirements have led to development of packages based on complex formulations and film structures. This presentation will translate complex packaging performance requirements into key functionalities needed in the package and the process followed to select appropriate raw materials and package design to help simplify package development.

513786 Upgrading a ZINC OXIDE Process to Continuous Vacuum Filtration from Batch Filter Presses

Abstract:

Madison Industries (Madison) and BHS-Sonthofen Inc. (BHS) have worked together through laboratory testing and field pilot testing to develop an optimized filtration process for a zinc oxide product. The current process uses a plate and frame filter press for the filtration of a zinc oxide slurry made up from a mix of various zinc material sources. The Madison objectives were to expand the production process and replace the present labor intensive process with a continuous operation. Creativity and â€œoutside-the-boxâ€ thinking were critical as Madison and BHS engineers looked at the process from new vantage points to replace existing filter presses. Continuous vacuum filtration was selected to provide maximum filtration efficiency and improve product quality while increasing yield and reducing operating and maintenance costs. With the installation of the vacuum belt filter in 2016, Madison has realized a 50% savings in wash liquids per batch as well as a reduction in labor and operating costs as the vacuum belt filter is a fully-automatic operation. The Madison and BHS collaboration illustrates a successful relationship between client and technology supplier. The BHS approach of lab testing, pilot testing, and idea-generation provides clients with optimized process solutions for critical and difficult solid-liquid separations.

512172 Data-Driven, Model-Based Approaches to the Scale-up, Transfer and Alignment of Processes

Abstract:

A very powerful approach to the scale-up of processes based on models built from data collected on processes at the different scales is presented and illustrated with several industrial examples. The technology is also applicable to the transfer of production between manufacturing sites, and to the alignment of existing plants producing similar slates of products. If the scale-up, transfer or alignment involves processes that are almost identical in terms of design and sensors, the technology is quite straightforward. However, the technology is most powerful when the scale-up, transfer or alignment is between processes of different design with different sensors and actuators and yet requires all the processes to manufacture a similar slate of products or modifications of them.

Although the models are built using data from the processes at the different scales using empirical multivariate latent variable methods, theoretical calculations involving engineering calculations on mixing characteristics, energy balances, etc. are readily included.

Examples given will include: (i) the scale-up of a roller compaction process from a laboratory compactor to a full-scale compactor in a pharmaceutical company; (ii) the scale-up of a batch pulp digester process; (iii) the transfer of production between batch reactors in a specialty chemical process; and (iv) the alignment of batch reactors in a biopharmaceutical plant to achieve similar yields and quality.

512407 How to Successfully Scale-up Batch Reactors

Abstract:

Process development is now increasingly based on the use of simulation software that enables chemical engineers to understand the characteristics of these processes and define the possible areas of improvement in terms of operational safety, product quality and operating costs, on all phases of scale-up: from the laboratory to the pilot plant and then to the industrial unit. In processes using discontinuous reactors (or fermenters), extrapolation is a particularly complex stage in the development phases. Process extrapolation techniques are based on the principle of similarity (geometric, kinematic, dynamic, chemical and thermal). The implementation of these principles in a specialized software makes it possible to determine, from the exploitation of the data collected at the laboratory stage, the characteristics of the reactor on a larger scale. The rigorous modeling of the equipment thus determined allows the process engineer to refine the sizing carried out. This methodology will be presented on various examples, particularly relating to heat transfer in batch reactors. Indeed, only a good control of the temperature throughout the reaction process, makes it possible to guarantee the quality of the products and in many cases the safety of the operation. However, the cooling (or heating) of the reactor will evolve throughout the production batch because the three parameters governing the heat exchange will evolve: the exchange area (since the level in the tank is changing), the temperature difference (since the temperature in the reactor is changing) as well as the heat exchange coefficient (since the properties of the process fluid evolve). All these parameters must be correctly taken into account in the reactor model so that the designer can rely on accurate simulation results.

510542 How Proper Startup and Shutdown Procedures Are Critical in Operating the Laboratory Fixed Bed Reactors

Abstract:

Laboratory scale fixed bed reactors are widely used for testing novel catalysts for chemical processes. These reactors are very small with reactor diameter, typically, in the range of ¼" to ½" and hold about 5-25 grams of catalyst. They are ideal for initial screening of many different catalyst samples at a very small scale. These reactors must be set up properly to ensure good contacting of the reactants with the catalyst particles. Good procedures for startup and shutdown of the reactors are equally important and can play a critical role in the reactor's performance. A real life example will be shared showing how a reactor's performance improved dramatically when the reactor's startup and shutdown procedures were changed.

513821 Designing Carbon-Based Materials for Efficient Hydrocarbon Synthesis

Abstract:

Electrochemical reduction of CO_2 (CO_2RR) using renewable electricity has emerged as an attractive approach to synthesize hydrocarbons and mitigate atmospheric CO_2 concentration. Carbon-based materials have recently attracted a particular attention as electrocatalysts for CO_2RR due to their extraordinary properties such as low cost and flexibility in tuning the morphology and electronic structure. Wide range of different N-doped carbon materials have been experimentally tested and reported for CO_2RR. The performance map for CO_2RR over reported N-doped carbon-based materials presents diverse activities and in most cases the major product of the CO_2 reduction is found to be CO. Based on a simple thermochemical analysis Wu et al previously suggested pyridinic-N as the active sites.[1] Using density functional theory (DFT) calculations we study a wide range of different possible active motifs included in the defective carbon materials to rationalize the associated trends in their intrinsic activities. We study both CO_2RR and hydrogen evolution reaction (HER) using thermochemical analysis. We identify several active motifs that are expected to exhibit a comparable or even better activity to the state-of-the-art gold catalyst, and several configurations are suggested to be selective for CO_2RR over HER. This study provides insights in the nature of active and selective motifs for CO_2RR and rationalizes a number of reports in the literature.[2] We also show that in the presence of metal impurities further reduced products can be synthesized with significantly improved activity and selectivity.[3]

References:

(1) Wu, J.; Yadav, R. M.; Liu, M.; Sharma, P. P.; Tiwary, C. S.; Ma, L.; Zou, X.; Zhou, X.-D.; Yakobson, B. I.; Lou, J.; Ajayan, P. M. *ACS Nano* **2015**, *9* (5), 5364â€“5371.

(2) Siahrostami, S.; Jiang, K.; Karamad, M.; Chan, K.; Wang, H.; Nørskov, J. *ACS Sustain. Chem. Eng.* **2017**, acssuschemeng.7b03031.

(3) Jiang, K.; **Siahrostami, S.**; Li, Y.; Lu, Z.; Gardener, J.; Lattimer, J; Stokes, C.; Hill, W.; Bell, D.; Chan, K.; Nørskov, J.K.; Yi Cui, Y.; Wang, H. *Chem* **2017** https://doi.org/10.1016/j.chempr.2017.09.014.

513277 Leveraging Separations to Improve Sustainability of Chemical Processes

Abstract:

Typically, separations technologies for chemical processes are designed to ensure that the desired product(s) meet technical requirements such as purity and that the overall process economics are optimized. We can improve the sustainability of the chemical products by also considering separations that improve energy efficiency, reduce waste and avoid toxic solvents, entrainers, and adsorbents. This presentation will use real world examples to review developments in this area and outline strategies to select and design separations technologies for more sustainable chemical products.

513173 Soot Generation of Olefin and Alkane Flares

Abstract:

Soot Generation of olefin and alkane Flares

Anan Wang[1], Helen H. Lou[1*], Daniel Chen [1], XianChangLi [2], Christopher B. Martin[3]

(1)Dan F. Smith Department of Chemical Engineering, Lamar University, Beaumont, TX, (2)Mechanical, Lamar University, Beaumont, TX, (3)Chemistry and Biochemistry, Lamar University, Beaumont,

Abstract

Flares are used to combust unwanted natural gas or hydrocarbon by-products of petroleum refining and chemical manufacturing. Recently, flare operations have been scrutinized due to their potential environmental impacts. Industrial flares may generate black carbon i.e. soot. However, the prediction of soot generation from flares is tricky due to fundamental combustion chemistry.

In this work, detail comparison of soot precursor species and soot yields between olefin and alkane flares were conducted based on the combustion chemistry. Optimal settings of controlled variables, including flow rate of steam/air assist and flow rate of make-up fuel, were identified in order to minimize soot yields.

513109 Modular Coal Gasification Systems: A Fit for Future of Power Generation?

Abstract:

This paper explores the addition of gasifiers to coal-fired power plants to provide a flexible technology option for the accommodation of renewable power that is only available intermittently, and is the primary source of base-load coal fired plants being required to cycle. Modular gasifiers could enable leap-ahead strategies for power plant operations, and for emerging economies, by providing an integrated modularized coal utilization technologies that can define a power plant of the future.

Coal gasification is very amenable to the modular concept, whereby coal is converted into syngas or hydrogen for the production of electricity, fuels, and chemicals in small modular systems. These modular units provide a flexible technology option for using domestically available resources while meeting future environmental emission standards. In addition to efficiently producing electric power, a wide range of transportation fuels and chemicals can be produced from the cleaned syngas, thereby providing the flexibility needed to capitalize on the changing economic market. As a result, gasification provides a flexible technology option for using domestically available resources while meeting future environmental emission standards. Poly-generation plants that produce multiple products are uniquely possible with gasification technologies.

Meanwhile, coal-fired power generating units in the U.S. continue to switch from high-capacity-factor operation to various modes of flexible operation. Cycling and load following are not new. However, over the past decade, and in particular the past few years, a perfect storm of factors has gathered to require even baseload coal plants to prepare for increased flexible operation.

Integration of the gasifier with the coal fired units is possible because of the many advances we have made in modeling, materials research, and advanced manufacturing. New developments in Computational Fluid Dynamics (CFD) modeling should be used to create new gasifier designs. These new designs for coal gasification systems which incorporate modular concepts will allow the gasifier to be easily added, expanded, or replaced as necessary. New materials that are being designed can be coupled with 3-D manufacturing can be used only where needed, for example in high temperature regions of the gasifier, to minimize costs.

510750 Towards Standardisation of Process Intensification in Solids-Handling Processes – the Ibd Platform

Abstract:

Process Intensification (PI) is broadly defined as *any chemical* (or process) *engineering development that leads to substantially smaller, cleaner, safer and more energy efficient technologies* and is achieved by significant enhancement of transport rates. To date, the majority of achievements in process intensification have been in liquid and/or gaseous phase applications.

Intensification of processes in which solids are an intrinsic component is significantly more difficult, with fouling an inherent problem which is contradictory to one of the key PI aims of equipment: miniaturisation. However, there are various examples of successful PI technologies across the broad spectrum of solids handling, including but not limited to the Taylor-Couette reactor, the twin-screw granulator and the toroidal fluidised bed.

Intensified by Design (IbD) is a large, multi-partner project funded by the European Union Horizon 2020 programme (SPIRE-08-2015, Grant No. 680565) which brings together a team of experts in PI, fouling, process analytics, process modelling and process control from across Europe. The main project output is the IbD Platform, a holistic, computer-based methodology for undertaking process intensification in solids handling processes. Features of the IbD Platform include:

- A comprehensive database of existing PI technologies and a knowledge-based selection methodology for initial shortlisting of options.
- A number of "*built-in-modules*" giving comprehensive design procedures for technologies from the PI database.
- A PI-based TRIZ methodology to guide new process innovations.
- Strategies for fouling prevention and remediation.
- Strategies and tools for advanced monitoring and control.

The IbD Platform and methodologies are validated against 6 industrial case studies from a range of sectors: mineral processing, ceramics processing, pharmaceuticals and metallic nano-particle production.

513010 Process Design and Development of Offshore NGL (Natural Gas Liquid) Recovery Processes

Abstract:

Process Design and Development of Offshore NGL (Natural Gas Liquid) Recovery Processes

Sekwang Yoon[1], Jin-Kuk Kim[1], Chi-Kyun Park[2], Byung-Don Lee[2]

[1] Department of Chemical Engineering, Hanyang University, Wangsimni-ro 222,

Seongdong-gu, Seoul, 04763, Republic of Korea;

[2] Institute of Chemical & Environmental Process, Jeonjin Entech Co., Ltd, Gadal 1-Ro 38, Gangseo-Gu, Busan, 46729, Republic of Korea;

This work focuses on process design and development for the recovery of ethane (or propane) from natural gas liquids (NGL) to be applicable in the offshore environment. In this work, the NGL recovery is mainly made through the distillation at sub-ambient conditions, and low temperature condition is created by the use of a turbo-expander as well as J-T valves, which are interconnected with a demethanizer column.

Considerable efforts have been made in process design to improve energy efficiency of NGL recovery process without compromising product specifications. First, heat integration techniques have been applied to maximize heat recovery existed within the process. Second, attempts have been made to fully realize any potentials for power recovery. Third, column analysis has been carried out to identify any inefficient features in column design, and column is designed in an energy-efficient manner by introducing a few side-reboilers and selecting optimal flowrate and operating conditions for feed streams. Finally, optimization method has been rigorously applied to determine the most appropriate configurational features (i.e. interconnections between a column, heat exchangers and a turbo-expander.) as well as optimal operating conditions.

Offshore characteristics have been incorporated in the design of demethanizer column in this study, in which techno-economic impacts caused from ship motions are systematically considered. This work will also address control and operability issues, in the context of maintaining design feasibility of NGL recovery process, and achieving reliable and safe operation in industrial practices.

*This research was supported from the Technology Innovation

(10042424, Floating Production Platform Topside Systems and Equipment Development) funded By the Ministry of Trade, Industry & Energy (MI, Korea).

Korea Institute for Advancement of Technology(KIAT) through the Encouragement Program for The Industries of Economic Cooperation Region)(R0003996)

513304 *Thermodynamic Data Requirements for Biodiesel Process Development*

Abstract:

Biodiesel has been the drop in solution for the renewable energy production particularly from various waste sources. Many new technologies are emerging with data demands for efficient scale-up and process design. The presentation will highlight the data requirements, gaps in the thermophysical property data, and consolidation of experimental and estimation methods for effective process development. A case study on how new calorimetric studies for heat of mixing measurement and Aspen flowsheet simulator were utilized for development of a 100,000 gallon/year continuous mobile mini biodiesel plant.

513337 Enzymatic Reaction Kinetics for Biodiesel Process Development and Scale-up

Abstract:

Biodiesel production from waste resources has been widely practiced as a possible solution to the increasing demand for renewable energy. Many new technologies for this production have been developed and require experimental data for efficient process design and scale-up. One such technology is the use of immobilized enzymes for biodiesel production. This presentation highlights the required kinetic data for effective process development of enzymatic biodiesel production using process simulation software. Isothermal calorimetric experiments and an Aspen process simulator were utilized for the development of a 100,000 gallon/year continuous miniature biodiesel plant.

513405 Direct Process for Methylchlorosilanes: Chemistry and Technology

Abstract:

Dow Performance Silicones (DPS) is the world leader in the direct process technology for the production of methylchlorosilanes (MCS, Me_xSiCl_y). MCS is the intermediate product to produce Poly Dimethyl Siloxanes (PDMS) that is the core base material to produce several thousand final silicone polymers. Silicones are synthetic polymers that combine the characteristics of plastics and glass with mixed properties of organic and inorganic polymers. MCS are produced *via* the Direct Process reaction of silicon metal with methyl chloride gas in the presence of copper catalysts and other metal promoters such as Sn, P, and Zn. Direct process is a unique solid-gas heterogeneous catalytic reaction that is more efficient with a bulk metal catalyst than a supported metal catalyst. The direct process catalysis involves generation of a consumable reactive site that repeats the cycle *via* regenerating the catalyst. This is unlike the *active site* regeneration in traditional heterogeneous catalytic processes. Reaction deactivates by the accumulation of impurity metals on the surface and methyl group decomposition that cause diffusion and kinetic hindrance as well as surface coking. This poster intends to highlight the key catalysis mechanism and technology aspects of this major commercial process and challenges and opportunities that exist in process improvement.

514278 Development and Application of Simulation Models for Improvement of Highly Non-Ideal and Coupled Distillation Systems

Abstract:

An organic acid-water mixture is separated using an entrainer which forms an azeotrope with water. The mixture also contains smaller amounts of other components which are separated in downstream distillation columns. The system is highly non-ideal with significant recycle between these columns.

After the startup, desired separation was not achieved resulting in the loss of organic acid. A simulation model was developed to represent this system consisting of 5 coupled columns. Several case studies were run to understand the impact of impurities and other operating parameters. These case studies helped to stabilize and improve the performance of the columns.

512153 Rapid Development of New Product Formulations through the Efficient Utilization of Company Data

Abstract:

The continual development of new materials or products is the lifeblood of most chemical companies. Companies are constantly trying to: (i) develop products with novel properties for use in new applications, (ii) improve the quality and cost of existing products, (iii) reformulate products using new materials, and (iv) consolidate product formulations using fewer starting materials. However, their general success in these endeavors has been poor. In a recent Accenture benchmarking report (2015) entitled "Why Low-Risk Innovation Is Costly" (http://www.accenture.com/us-en/Pages/insight-low-risk-innovation-costly.aspx), executives indicated that indeed innovation is extremely important. 93% of executives surveyed said their companies depend on innovation for long-term success, but only 18% believed that their company derived a competitive advantage from its innovation strategy, and most of the executives were less satisfied with their company's current innovation performance (2015) than they were in 2009 (previous benchmarking report).

Product development at most big companies is characterized by:

- A lot of trial and error and intuition-based experimentation
- Some use of DOE's – but usually only in a very local manner – no global overview
- No repository for the accumulated product development knowledge of the company
    - No easily accessible database of all past development data
    - No good database on raw material properties
    - No repository for past modeling efforts
    - No good way of sharing information among developers
        - Intuitive experience of individual researchers is often only way they carry knowledge forward.
        - Little of that information is effectively transferred to other developers.
    - No effective ways of rapidly finding formulations involving new raw materials
    - Processing conditions are often not included in development studies, but left to the scale-up group

There are basically 3 general degrees of freedom in developing, improving, or reformulating products: (i) the choice of raw materials or ingredients, (ii) the ratios in which to combine or react them (formulation or recipe) and (iii) the process conditions under which to manufacture them. These three degrees of freedom are highly interactive, making it important to consider them simultaneously.

In this presentation we discuss powerful multivariate latent variable data analysis methods for extracting information from diverse development databases (raw materials, processes, formulations and quality), for optimally augmenting the databases with multivariate DOE's, and for using that information via optimization to rapidly formulate products that meet all the desired quality specifications at minimum cost.

A number of past development projects will be used to illustrate the approach. These include the development of novel functional polymers for use in medical devices and for the cores of the very popular Srixon golf balls (used by a large number of the touring pros), the development of novel high performance polymeric coatings, and the development of new food formulations.

Although this presentation will refer to advanced multivariate data analysis, design of experiments, and optimization, it will not be essential for the attendees to have a detailed knowledge of these in order to understand the essential ideas and the potential of the methods. The emphasis will be on the basic problem of rapid product development through the efficient quantitative utilization of company data, and on illustrating this approach with successful industrial examples.

513514 Innovating Innovation Processes for Chemical R&D By Treating As Metaphorical Chemical Processes

Abstract:

Both industrial and academic research produce innovations that can be used to add value to products and services. Many labs seek ways to produce innovations that 1) have higher value, 2) are faster to market, and 3) are produced as a consistent stream. We have found that by treating the innovation process itself like a chemical process, the flowrate of valuable innovations can be increased. In this talk we analyze the innovation process according to streams (e.g., desired outputs, inputs, recycles, purges) and unit operations (e.g., idea generation "reactors" and idea evaluation "separators"). This perspective enables the analysis and measurement of R&D, providing a coherent philosophy for innovation processes.

513106 An Experimental and Computational Study of the Four Blade Loop Impeller for Liquid Dispersion during Suspension Polymerization

Abstract:

The current work describes the characteristics of a Four Blade Loop Impeller (FBLI), which is intended for applications such as suspension polymerization, wherein a narrow range of suspended droplet diameters is desired. Using a dilute dispersion of silicone oil in water, an experimental measurement of the droplet size is performed using Particle Vision Microscopy (PVM) technique. Correlations between the maximum droplet chord length and maximum energy dissipation rate in the system are reported, in agreement with literature. The FBLI shows a narrower range of droplet sizes compared to a standard Pitched Blade turbine at the same impeller based Reynolds number. Computational Fluid Dynamics (CFD) simulations of this laboratory scale system with this impeller and other canonical paddle impellers showed a narrow rate of shear on the impeller surface for the FBLI. The CFD model was validated with experimental results from Particle Image Velocimetry measurements, and shaft power measurement. Power and flow numbers of the impeller are reported.

DIFREX[R] Nexgen Expands and Advances Reactor & Technology Solutions

Subhash Dutta, Stephen Arnold, Aashish Gaurav, Jim Brenner
DIFREX LLC, 8900 Astronaut Blvd. Cape Canaveral, FL 32920 USA

Chemical reactors are the most vital elements of the chemical industry, contributing significantly to the total plant cost, efficiency of production, pollution control, number of separation steps needed downstream, and therefore to its ultimate profitability. Although the reactor is the heart of most process plants, it usually is treated as a "black box" or a proprietary item and is not covered by commercial simulators. Each technology developer or licensor uses its own procedure to develop its reactor model. Such a procedure often is lengthy and expensive, due to ill-defined steps, many trial-and-error mistakes, and excessive pilot-plant campaigns. Finally, even if the reactor is "successfully" scaled up to commercial size, the credibility of the design and the optimum operating conditions of the reactor often are questionable. "On the other hand, a plant owner may not have the know-how or confidence to modify, revamp, or modernize the reactor or its operation. The owner needs to build an in-house model for any such effort"[1]. The cover article "Build Robust Reactor Models" in the October 2000 issue of Chemical Engineering Progress presented a "well developed and tested procedure for building a robust reactor model/ model package, [with] tips and traps [plus] proven ways to substantially cut the cost and time for the effort"[1].

One of the authors of that article (Subhash Dutta) has spent his career on this work, plus other activities successfully "serving clients worldwide with the fastest, assured and most comprehensive reactor services and solutions in industry, to minimize risk and cost of process development; and to provide help in starting and making confident decisions on new reactor/technology or change/revamp of existing ones"[2]. Dutta's work has continued and advanced this work, including the General Reactor Model (GRM™) referenced in the article.

In addition to his further assistance to clients, he has formed a company, DIFREX[R] LLC, which uses the GRM™ and also the wide range of know-how and expertise of an expanded team. Today DIFREX[R] LLC is a team of process engineers based in Cape Canaveral, Florida, that provides reactor services and solutions to clients worldwide in areas such as design and consulting on new/existing reactor systems, critical review/quick-check of third-party designs, experimental development and validation of proposed and recommended designs, debottlenecking/retrofit/revamp, and emergency response[2].

This paper will provide more description of the GRM™ and the other services that DIFREX[R] provides, including examples of past applications and references[1-34] and how they have contributed to faster development, evaluation, decision-making, etc., reducing costs and time, and have also led to increased standardization and enhanced certainty of the results, with additional economic benefits. These advances have been applied to a wide variety of reactor systems, with or without catalyst, and will be extended to more in the future.

1. Who says that a Simple Fixed Bed is the Right Choice?

Packed/Fixed Bed Reactors (PKB/FXB), which employ larger particles in a static bed, are traditionally viewed as the most *obvious or first choices* in industry partly due to the traditional notion that they are the easiest to scale up. In fact, the scale-up of a multi-tubular fixed bed (**TFXB**) system is not a scale-up at all. A commercial reactor of this system simply uses a multiple of the successfully performance-demonstrated single catalyst-filled tube to reach the commercial production goal. However, this class of reactors suffers from many serious drawbacks that include the following: a) high heat and mass transfer resistances leading to a steep axial temperature profile in the reactor, b) poor catalyst utilization or low catalyst effectiveness factor, due to its use of relatively large particles, c) possible maldistribution of gas between all tubes in multi-tubular design (including impact of catalyst not loaded equally), d) no continuous solid (catalyst or material) regeneration and replacement capability, e) no load-following capability, f) hot spot formations possibly leading to catalyst activity/selectivity loss and even catalyst/material vitrification, g) higher pressure drop, and h) restricted window of allowable feed composition, temperature and/or pressure to avoid possible temperature runaway for highly exothermic reactions.

However, there are many alternative reactor designs, as illustrated in **Figure 1**.

Fluidized Bed Reactors, both Bubbling/Turbulent (BFB) and circulating (CFB), do not suffer from the drawbacks of PKB/FXB reactors indicated above. The smaller particle sizes in these particulate systems lead to better heat and mass transfer rates, allowing operation at uniform bed temperature, which is normally preferred. Close to 100% catalyst utilization can be achieved, with no gas maldistribution. These reactors also provide continuous catalyst/material regeneration and replacement capability due to fluidizability of the chosen particulates that can be continuously discharged and recharged after regeneration and/or feeding a make-up stream. These reactors also have load-following capability, allowing operation across a broad window of production capacity as desired. They can be operated with lower pressure drops and are usually also free of hot spots. The windows of allowable feed composition, pressure and temperature are much wider compared with a fixed bed reactor, with less chance of reactor temperature runaway. On the other hand, the key drawbacks of a fluidized bed reactor include the following: a) potential dust emission with possible environmental concern, b) loss of expensive catalyst by entrainment, to the extent that fines are not fully captured and returned, c) particulate attrition causing excessive fines buildup, leading to excessive solid loss or depletion of fines that are needed for good fluidization, and other potential problems like line plugging and agglomeration, d) potential erosion of lines or internal heat transfer tubes, e) poorer solid-gas mass and heat transfer rates for BFB through the interface films of gas bubbles or voids, and f) potential neglect of safety concerns (even unacceptable bypassing of safety instrumentation and procedures) if operators do not fully understand, and become complacent with respect to the high flame-arresting safety benefits of a BFB, e.g., for partial oxidation (POX) reactions, including knowing when or where they do not apply.

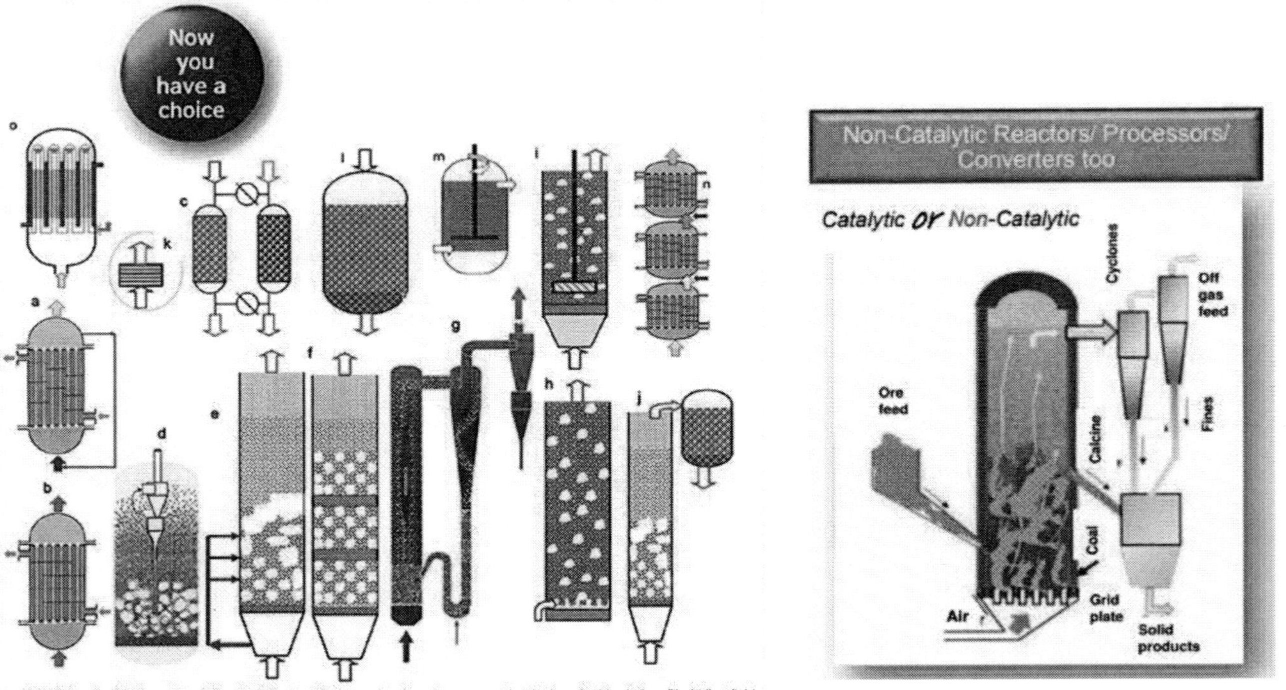

a, b)Tubular fixed bed reactor with and without product recycle: c)reactor-regenerator dual-packed-bed: d, e, f)bubbling fluid bed without and with quench/ inter-stage feed injection, and with redistributors: g)fast/ circulating fluid bed: h, i)fluidized bubble-column and stirred slurry reactor: j)hybrid (fluidized bed+polishing) reactor: k)micro-channel reactor. l)packed bed, m)stirred tank, n) quenched tubular fixed bed, o)double wall bayonet tube reactor

Figure 1: DIFREX technologies encompass numerous catalytic and non-catalytic reactors

Circulating Fluidized Bed Reactors (CFBs) are usually designed for fast reactions and for reactor-regenerator dual-bed systems. In the dual-bed system, the solid particulates, catalyst or non-catalytic material, can be regenerated in the regenerator bed, or sometimes in a transfer line, and be returned back to the reactor in a continuous circulation of the solid. (This is also done with some BFBs.) The primary section of a CFB system operates without gas bubbles or voids, and has minimal solid-gas heat and mass transfer resistances. However, it suffers from the following drawbacks compared with the other systems: a) bed voidage throughout these reactors is usually very high (in the range of 88-98%), thus, solid concentration per unit bed volume very low, resulting in much lower reaction rate per unit bed volume, b) these reactors operate at relatively much higher gas velocities and therefore are applicable to fast reactions only, c) at the high gas velocities, which includes the catalyst movement, they demand much tougher or more attrition-resistant catalyst or material unless the material is desired to be broken down (or reacted away) and to pass once-through as in gasifiers or combustors, d) the reactors need to be much taller (in the order of 30 meters or even more) for adequate reaction residence time to achieve satisfactory conversions at their high gas velocities.

Figure 2 illustrates the extensive and convenient framework and easy use of the DIFREX software as represented as a **dial-a-reactor**. It incorporates detailed flow and hydrodynamic behavior for many different reactor types, allowing prompt and easy calculation switching (dialing) from one reactor type to another, for the same chemistry and kinetics of the reaction system. Better, faster initial determination of the best conceptual/preliminary design for a reaction system, catalyst or process can therefore ***now*** be achieved by comparing the design and performance details projected by the software at the earliest project stages. The best choice could be a packed bed (PKB), a tubular fixed bed (TFXB), a bubbling/turbulent fluid (BFB), a circulating fluid bed (CFB), all with or without quench and/or recycle, or a hybrid reactor (CSTR-CFB-BFB) as in a typical FCC reactor or other system. The best **Reactor Of Choice (ROC)** can now be determined along with the **Catalyst or Material Of Choice (COC/MOC)** and much earlier – even while the solid catalyst or material is being developed. Final determination of best ROC,

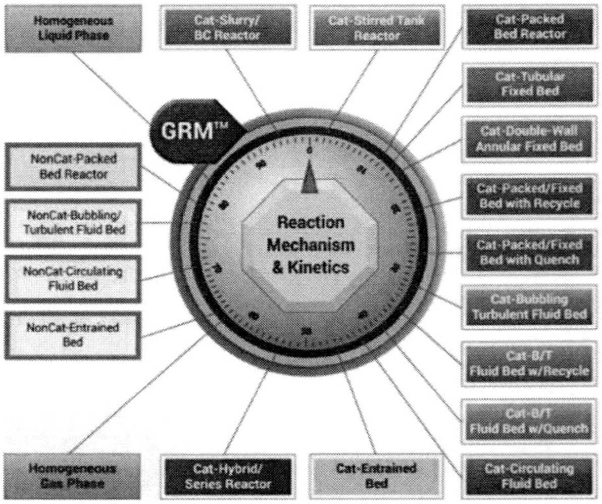

Figure 2: Extensive and convenient framework for reactor selection

however, would be determined by **consideration of other criteria** that include cost, particulate handling/recycle, strength and attrition resistance of solid catalyst, and miscellaneous other characteristics like particulate fluidizability and agglomeration problems for fluidized bed systems. However, the software allows one to reach each ***GO/NO-GO decision gate*** much sooner, leading to substantial savings in both cost and time for catalyst, process or technology development.

Current DIFREX team members were instrumental in successful commercialization of several new process technologies, involving all four basic solid-gas reactor types: PKB, TFXB, BFB and CFB reactors and their variations. A prime example is the ALMA process, a fluidized bed POX process for production of maleic anhydride from n-butane and air. These designs were products of seamless integration of particle technology (including particulate flow and hydrodynamic behaviors) with chemical reaction engineering principles. Results of some of these works are available in a series of publications by Dutta, Arnold, and other authors[1-20]. Many of the theories and procedures used in these works can be found in the two classical books on Fluidization Engineering and Chemical Reaction Engineering by Kunii and Levenspiel[21-22].

Scale-Up Fear: No More

Due to significant advances made in both chemical reaction engineering and particulate system design and engineering, and successful demonstrations over the past three decades in commercial processes, there is little reason any more to fear scale-up failure for most of the reactor types used in the industry today. Failures and/or delays still seen are primarily due to questionable or random (non-standardized) kinetic data generation procedures that lead to unreliable data that are not suitable for successful scale-up.

2. Who says that Plant Expansion or Upgrade/Change is Difficult?

DIFREX software, services and solutions integrate chemical reaction engineering with particle technology to provide the right solutions for design, development, scale-up and successful commercialization of gas-solid reactors, enabling one to promptly establish the design and performance of virtually any particle or solid-gas reactor type for reaction systems of different complexities. Since it has fundamental basis in the two engineering sciences (reaction and particle) and avoids much of the empiricism of traditional process development, especially in the HC processing industry, it is better suited for new designs and also revamp or upgrade of an existing technology/ reactor/ process, particularly for a new feed or product stream. This refers to both catalytic and non-catalytic solid-gas reactions, including the use of a catalyst (active components with or without carrier) as the solid for catalytic processes or a non-catalytic solid particle like coal, biomass or a mineral (FeO or NiO, for example) for combustors, gasifiers, and metal extraction/roasting processes, and also solid particles as adsorbents to transfer atoms and other molecular species from one stream or vessel to another.

Reach *Decision Gate* Quickly

*How **not** to see that moth-balled R&D/ Pilot Plant/ Demo Campaign Again?* Whether it is a new venture or a retrofit, revamp or upgrade, delay and cost overrun are common even to reach the first ***Go/No-Go decision gate***. This adds to the pervasive fear of unmatched (below promised/projected) reactor performance, or worse, total failure. We have witnessed these events in multiple ventures that succeeded or failed. We have critically examined and documented the causes and effects. We are now ready to offer solutions to minimize cost and time to reach the decision gates and then minimize fear and risk for the *final* development.

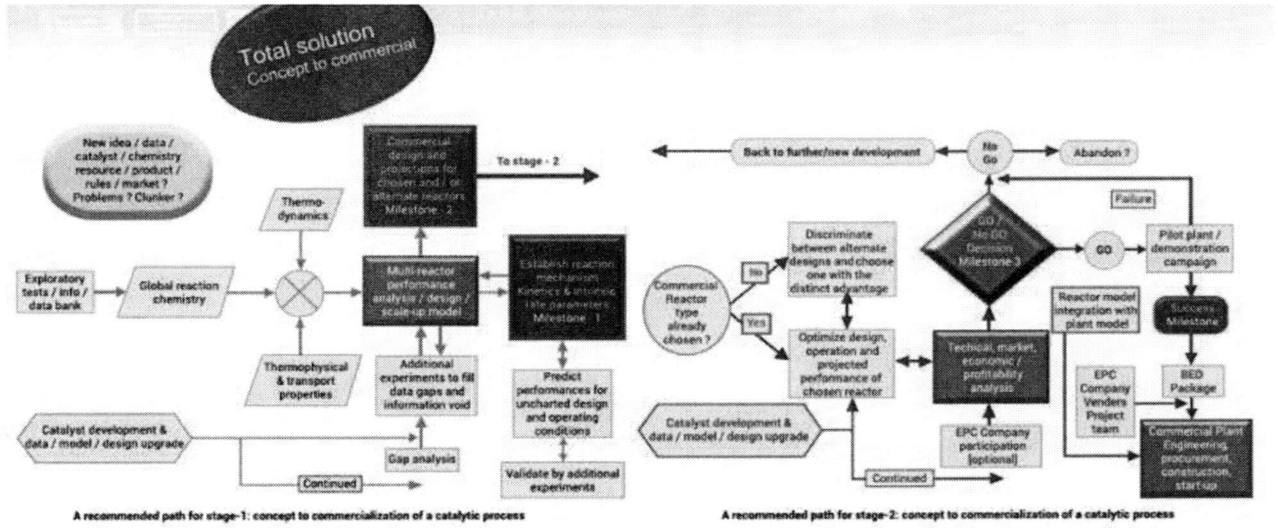

Figure 3: Concept to commercial action flowchart, critically examining reaction and process engineering at each step

A2Z: Fastest Path from Lab to Commercial

Due to combined background, experience and expertise in concept validation, reaction engineering, detailed design of reactors and auxiliary systems (like cyclones, diplegs, circulation loops, bed internals, bubble size control, distributor grids), detailed modeling and simulation, and process engineering, DIFREX can help in your next or ongoing venture. The service covers the entire process or ***total solution*** (**Figure 3**) from concept or lab to commercial, and accomplish the task successfully at the fastest speed possible.

Figure 4: GTL unit on a shipboard

Go Ahead

So, go ahead. • Add that sulfur guard/device for emission control or for feed gas treatment, • Place it at a well mouth or along distribution lines, • Install a flare/ waste stream converter for profit and environment, • Install a mercury emission control for your sludge/waste incinerator, • Install concentration monitoring controls to meet environmental regulations, • Install a GTL unit at a well mouth, remote island/location or on shipboard (**Figure 4**), • Carry out *retrofit, revamp or upgrade* you have postponed, • Revive troubled/idled units, • Minimize over- or under-designs.

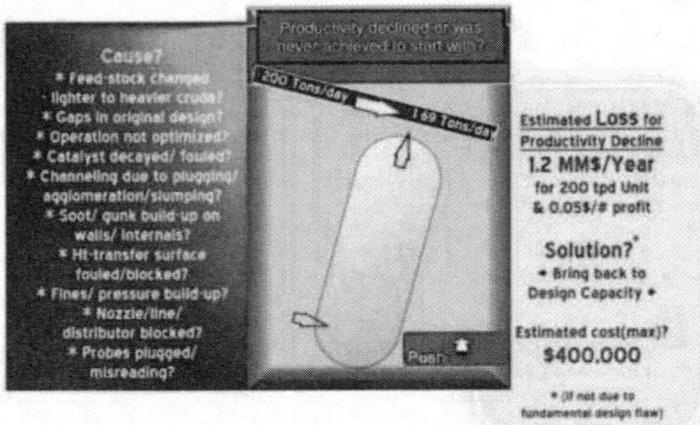

Figure 5: Robust reactor design can address problems in underperforming, uncontrollable, unpredictable reactor systems

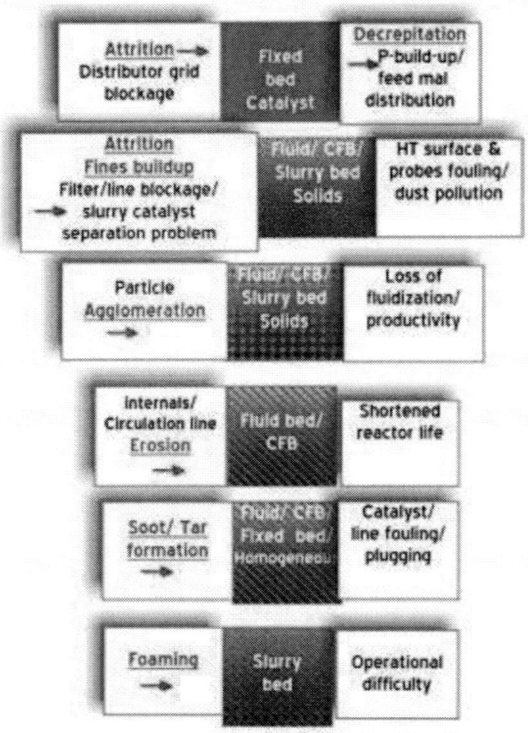

Figure 6: Prominent reactor design issues often ignored in most design software models

Take the fastest path to 3-D Reactor Solution → Design, Development and Debottlenecking

Underperforming, uncontrollable, unpredictable, problem-ridden or dying reactor? Or, one sitting on the side-line/off-line? Help is NOW available. Both on- and off-site plant unit debottlenecking on short notice and in-depth assessment of estimated benefits, cost and return on recommendations or optimization of running unit or retrofit/revamp is available. Most often, the benefits and quick returns would strongly justify the modification as shown schematically in **Figure 5**. Prompt diagnosis and solutions to troubled or idled reactors are *now* available.

One-stop solution

Although many details beyond the key design specs and operating conditions are to be filled by process engineers, there are issues that still fall under the primary responsibility of the reactor designer. Some of these issues are shown in **Figure 6**. These aspects are usually beyond the scope of most computer models. But ignoring these issues in advance, or inadequate/wrong solutions to these anticipated problems/issues may lead not only to project failure, but also to disastrous consequence in a running plant.

Reactor as "unit operation": Unlock true potentials plus easy revamps and upgrades – plant-wide

A reactor or reaction system (shown in **Figure 7**) is the heart of the plant, but it is often not treated as another unit operation in a process plant – like a separator, settler, heat exchanger, pump or compressor. It is usually treated as a **black box** where all upstream operations feed into and downstream operations begin. The rest of the plant design utilizing unit operations is constrained to meet the needs of the feed and narrow feed/operating conditions and to *treat the effluents* according to the original design of the reactor. Reasons include protections of trade secrets and control by reactor vendors, and presumed risk and fear.

Figure 7: Reactor as heart of the plant, but often treated as black box

DIFREX technologies simplify the design and development of reaction systems as well as debottlenecking operations. One can now unlock the full potentials of the plant a) by treating each reactor as another *unit operation* and its *true/full optimization* by going beyond the *narrow* boundaries of designs and operations – possibly *locked in* by the original design and pre-conception, b) optimized for *new/changed feedstock/* feedstock composition, and c) optimized for new/changed product/ product composition/ market demand. These operations can lead to savings in upstream and/or downstream processing costs due to improved/new reactor feed and effluent conditions or better yield. Although the original capital cost of a reactor may be only 10-30% of total plant capital cost, a substantial savings in capital and/or operating costs of the up- and downstream processing units by the new and true or full plant optimization process may lead to substantially improved profit from the operating plant as shown schematically in **Figure 8**.

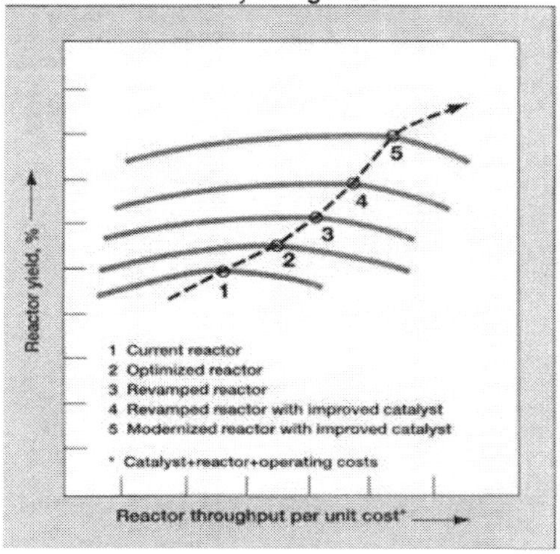

Figure 8: Reactor improvement steps and associated cost improvements

Plant Expansion or Upgrade/Change

Plant Expansion or Upgrade is *now* far easier and cost-effective, as illustrated in a series of three publications by Dutta and Gualy in Hydrocarbon Processing in 1999, titled "*General Reactor Model Improves HPI Applications*", "*Overhaul Process Reactors*", and "*Modernize Process Reactors*" [4-6], and in "*Super-modeling Reactors for Profitability*" by the same authors published in Chemical Engineering in year 2000[3]. An example is shown in **Figure 9**, where substantial benefits can be gained by moving from a fixed to a fluidized bed reactor. MTY in this figure stands for mass time yield, which is catalyst productivity in ton of product per day divided by ton of catalyst required. Another example is illustrated for a similar reaction system in **Figure 10**, for change from a near-atmospheric fluid bed (AFB) to a pressurized fluid bed (PFB) for substantial process benefits. Both changes were accomplished for maleic anhydride fluidized bed development and displacement of multitubular fixed-bed reactors for the ALMA Process, but with different actual parameter values and impacts than shown. Clearer views and explanations for these examples and more are available in the four cited references.

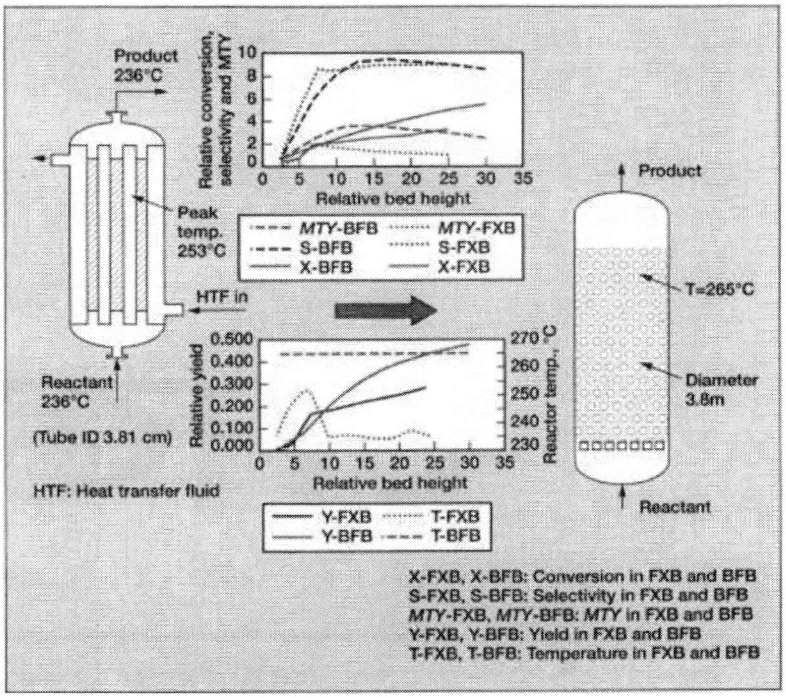

Figure 9: Expected benefits of a switch from Fixed Bed to Fluidized Bed reactor: illustrative example

For key hydrocarbon processing reactors, for example, for a typical FCC reactor, simulation and performance optimization/improvement can be more easily achieved by modeling it as a hybrid reactor system consisting of a CSTR at the bottom followed by a CFB and then a BFB on the top, as illustrated in **Figure 11**. Hydrocracking and isomerization reactions for heavy oil upgrade can be simulated by a reaction scheme shown in **Figure 12** using surrogate compounds. Detailed accounts of such treatment can be found on the DIFREX website[32] and in the references cited above.

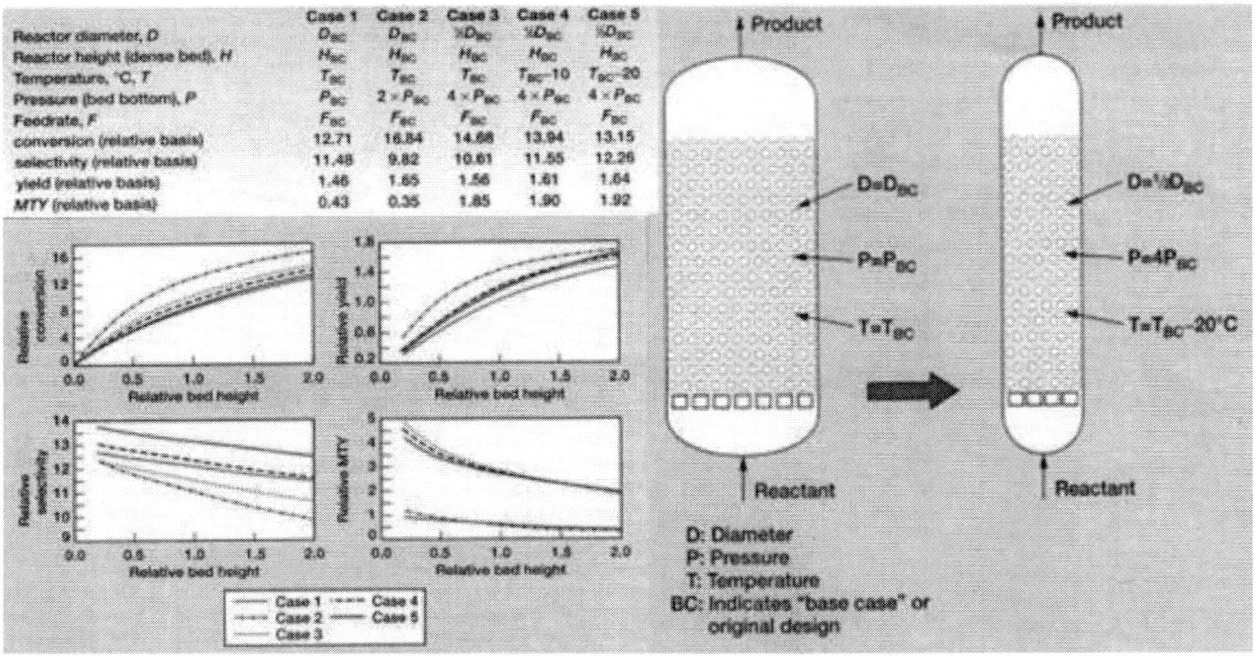

Figure 10: Switch from Atmospheric Fluid Bed to Pressurized Fluid Bed: illustrative example

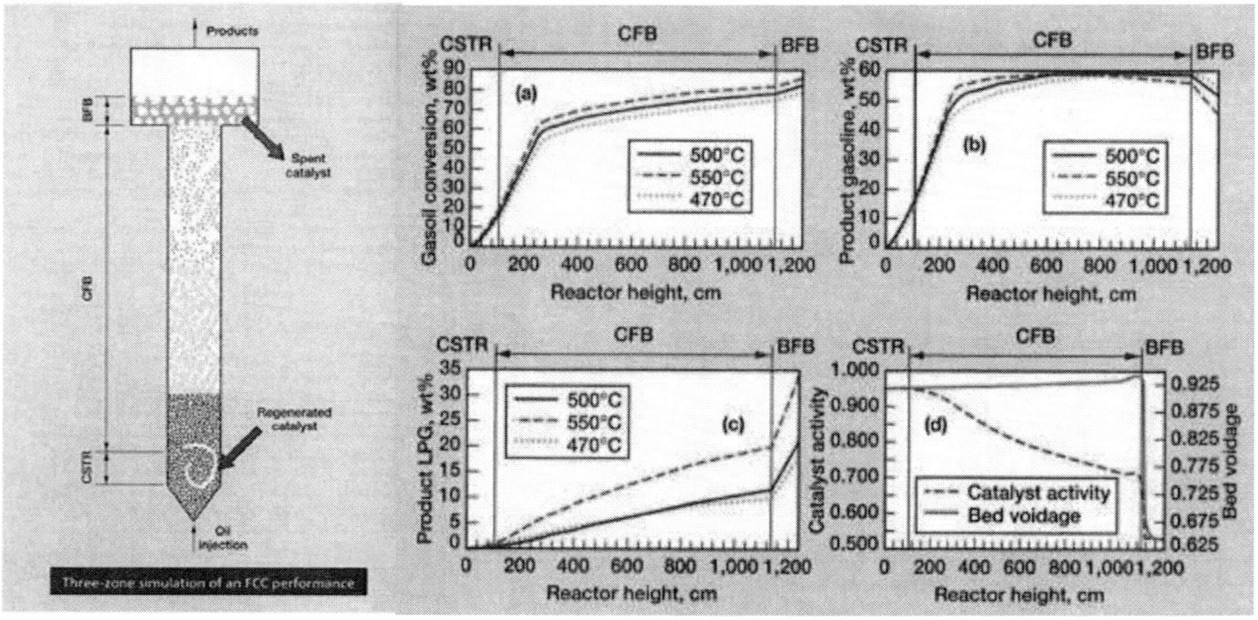

Figure 11: Performance Simulation of FCC with a simple 5-lump reaction model

3. Process Development: Randomness, Risk and Uncertainty

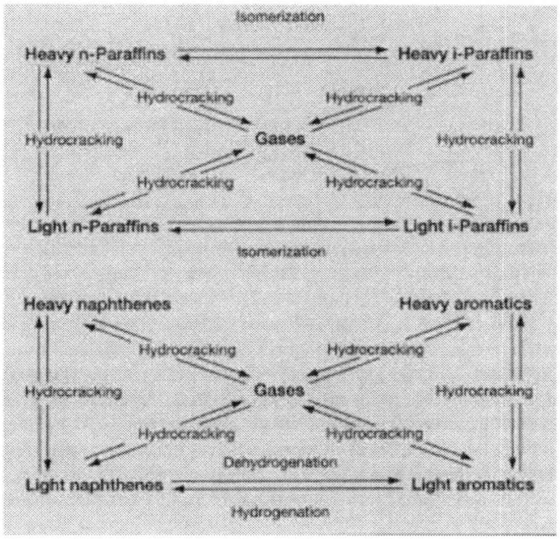

Figure 12: Typical Heavy Hydrocarbons Conversion reaction network

Standardized procedures: For head start and minimum delay and failure

No standardized procedures or steps currently exist for any of the following reactor tasks: • upgrade/revamp, • servicing, • data/performance analysis, • catalysis design/development, • optimization, • catalyst/process development, • scale-up/design, • modeling, • operation/control and • process innovation. The entire field is currently run by individual choice, conflicting opinions, randomness, and many trial-and-error steps. This causes cost overruns, delays, and risk in a new venture and, in many cases, project failures. Standardized procedures/steps in carrying out one or more of these tasks from concept to commercialization for a new technology/ catalyst/ reactor or for upgrade/ revamp of an existing or old technology/ catalyst/ reactor are now offered by DIFREX for head start and to minimize project delays and possible failures. It is accomplished by minimizing the randomness of the very process of development, design or redesign/revamp efforts of many projects that proceed like a runaway reaction and become uncontrollable. DIFREX offers a specific roadmap, schedule and milestone chart customized for each project or subproject, which help both in cost containment and timely and successful completion, including • catalyst formulation, • test rigs design, • test plan and matrix, • performance and data analysis, • kinetic and reactor model development, • safe reactor design, and beyond. The standardized procedure is applicable to a new technology/ catalyst/ reactor, an operating plant reactor, and also for using the reactor as another "unit operation" in the plant as shown schematically in **Figure 13** and powered by a friendly General Reactor Model software "smartpack" (GRM^{TM}). This is coupled with additional offers/advantages and coverage as shown in **Chart 1**.

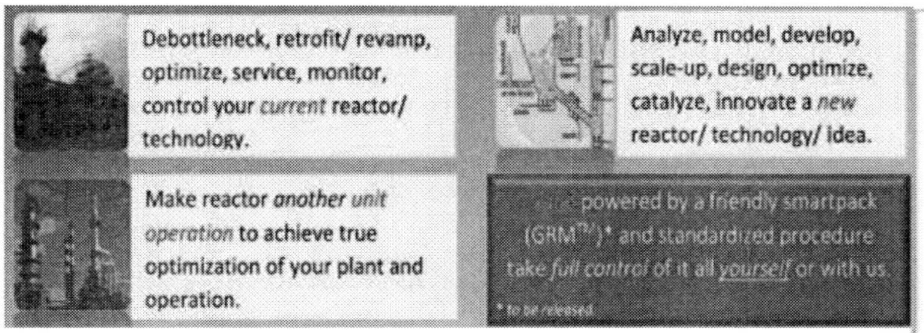

Figure 13: DIFREX offers standardized reactor design and implementation procedures

- *'No-Oops' A2Z solutions* NOW cover a *vista of reactors – reactors of virtually any kind* involving *reactions of any kind or complexity* and *even non-stoichiometric* reactions as in refineries, coal and biomass conversions.
- Tasks are *now* completed at a *fraction of time and cost* to reach *the Decision (Go-No Go) Gate*.
- Evaluate *alternate reactor concepts* at about same time to make *the right decision* on *ROC* (*Reactor of Choice*).
- *Head start* and *total CTC* (*concept-to-commercial*) services to *minimize failure*, and *improved reactor safety*.
- Aided by a *unique*, friendly, *secured and stand-alone* GRM™ software package (*smartpack*) that can hide in a wallet, and *standardized procedures*, tasks *now* become *far easier* than ever before.
- Building a *robust kinetic model* from *raw data* and minimum information that may be *key* to *commercial* success *and safe* operation is *now* made *easy*.
- Performance *guarantee* can be offered due to *our 'zero-failure' track record* of *multiple commercial successes* with fixed and fluidized and quenched *bed reactors* - backed by *our edge and insight* and advance 'oops' analysis.
- *Enterprise solutions* for sustained growth and edge.

Chart 1: Key/unique features of the services offered by DIFREX

Experimental Program

Misguided, lengthy and/or redundant and useless data generation often result from an experimental campaign for a new process development. DIFREX helps prioritize the areas of focus in recommending a test matrix that is conducive to collection of data most useful for promptly establishing reliable reaction mechanism and kinetics, catalyst screening, developing the reactor model, and successful scale-up and optimization of the commercial reactor design.

Kinetic Data, Temperature and Activation Energy (E)

DIFREX considers establishment of a reliable process/reaction kinetic model as the most important technical task for prompt and successful design, development or commercialization of a process. For catalytic gas-solid reactors, DIFREX recommends a specific (integral) test reactor design, plan and operating matrix for generation of *useful* kinetic data. Emphasis is placed primarily on measurement of the reactor axial temperature profile as accurately as possible, since the effect of temperature profile on the reactor or catalyst performance is usually far more than that of any other operating or design variable in most reactors. An accurate measurement of the temperature profile is necessary for evaluating the activation energy (E) of each reaction from the observed reactor performance data at various design and operating conditions. Relative values of the activation energies of reactions involved determine the product and by-product selectivities and yields for optimization for each reactor.

A reliable kinetic model must properly account for heat and/or mass transfer and other effects (shown schematically in **Figure 14**) that may mask the intrinsic reaction kinetics of a reaction/conversion process. Building a reliable kinetic model is a difficult and time-consuming task. DIFREX makes it easy even with minimal information in hand. DIFREX *has built reaction models as complex as in coal combustion, gasification and pyrolysis, hypergolic reactions of gel propellants in rocket engines and hydrocracking/hydrotreating of heavy oils, to as simple as ammonia and methanol syntheses.*

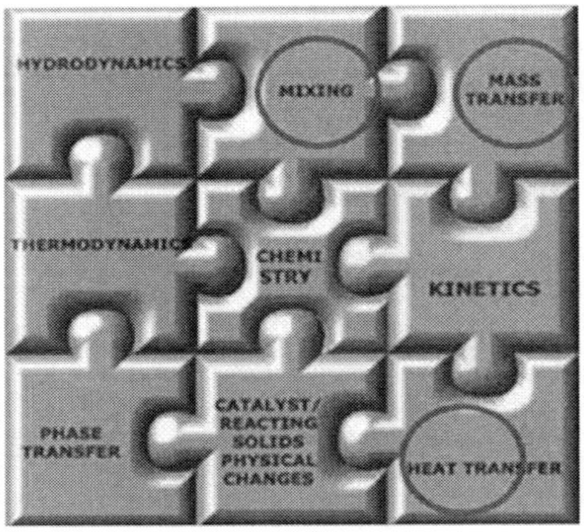

Figure 14: DIFREX kinetic models account for effects of other phenomena on intrinsic kinetics/process

Figure 15: Typical progression from concept to commercialization via DECISION gate

4. Why MUST a New Technology Venture be Expensive?

Speedy and satisfactory solutions need more than a math-pack

Judgment is needed in *discrimination* between alternative information sources, methodologies, procedures, correlations and databases to create *effective* and *easy* solutions with *adequate details*. Theories, correlations and procedures need to be tested and *re-validated* at every scale from micro- to small- to bench- to pilot- to commercial reactor/reaction systems of *every kind – to minimize risk* of scale-up failure or for evaluating/switching from one to another reactor system. DIFREX offers *zero-failure multi-technology commercialization track record* and extensive *experience* dedicated *exclusively to reactors of virtually every kind* and backed by coverage and track-record *unmatched* in the field. Also DIFREX design is built on *five focal points* – ***ease***, ***choice***, ***agility***, ***safety,*** performance ***ensurance***.

Time is everything

Now it is possible to move from concept to commercial (schematically shown in **Figure 15**) with confidence and at a much faster speed. The path includes catalyst/material development, determination of COC/MOC (Catalyst/Material Of Choice), testing and demonstration, modeling, design, scale-up, optimization, Go/No-Go decision, determination of ROC (Reactor Of Choice), process and detailed engineering, fabrication, construction/installation, commissioning and follow-up.

Both Catalytic and Non-Catalytic Reactor Designs

DIFREX's added strength is also in the design of non-catalytic (and mixed catalytic + non-catalytic) reactors wherein solids take part in both chemical and physical transformations. Areas covered include combustors/ incinerators/ gasifiers/ pyrolyzers of biomass/ sludge/ coal, minerals roasters/ calciners, desulfurizers and Hg removal units.

Design/ Development/ Revamp Done in 3 Quick and Easy Steps

It is accomplished by utilizing the GRM^{TM} Smartpack and following steps 1-2-3 (**Figure 16**). It is easy, prompt and user-friendly. Details are provided in references 32 and 33. Results for a comparison of design specifications and performances of three FT synthesis reaction systems are shown in **Fig. 17**, a quenched tubular fixed bed (QTFXB), tubular fixed bed (TFXB) and bubbling fluid bed (BFB). The *"Best Choice"* chosen here is from the comparison of the yields only. The final choice can be made by the user based on additional considerations that include mechanical/fabrication constraints, reactor cost, catalyst fines emission/management issues, attrition loss and circulation line erosion, among others.

The best Reactor Of Choice (ROC) for most systems can be determined ***within a week or two, if not in a couple of days,*** provided that and when the back-up test/ performance data of the reaction system are sufficiently complete and reliable. Furthermore, the DIFREX database on reaction mechanism and kinetics, thermochemical and transport properties, and methodologies can help quickly bridge the gap between available and necessary information and leapfrog a new technology venture to success.

Step 1

Collect system performance data on COC/MOC (Catalyst/ Material of Choice). Establish process mechanism and kinetics following the guideline provided. If full information on process mechanism and kinetics not available, the SmartPack built-in database should help you to fill in most of it for many systems. Preliminary design based on it should be adequate for most applications.

Step 2

Define your process by answering five (5) questions on this page:
1. Objective
2. Feed
3. Key Product/ Product Group or Process
4. Reactor of Choice from the list below
5. Scale – Bench, Pilot/Demo or Commercial

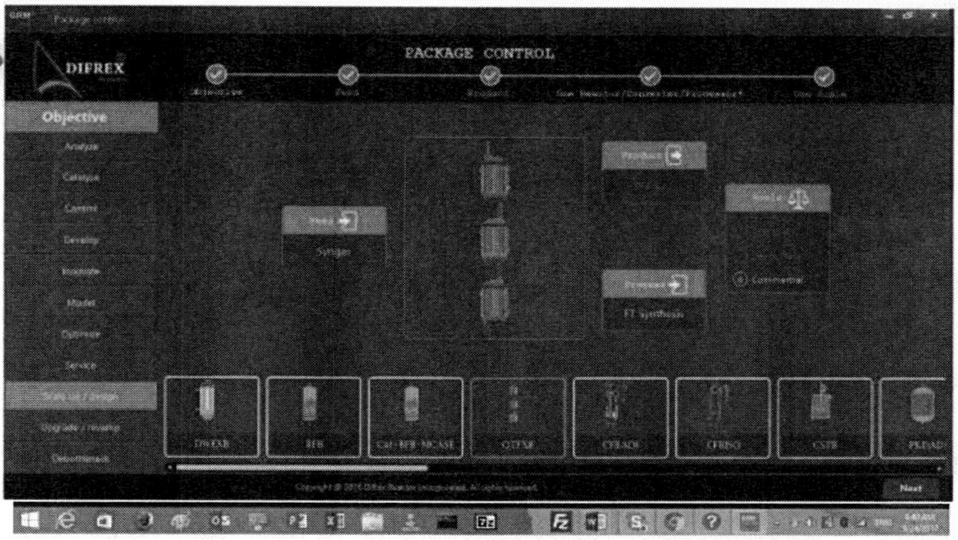

Step 3

Input* process mechanism/ kinetics, design/ operating conditions and catalyst/ material specs per top menu bar. Then:
1. Generate Input File
2. Execute
See input summary and results, showing complete design and performance summary in tabular and graphical forms and complete details@ in 'more >>' sheets.
- Complete case studies for the chosen Reactor by changing one or more variables on on-screen Input table.
- Compare cases on-screen.
- Repeat for another Reactor.
- Pick the Best or RIGHT reactor.

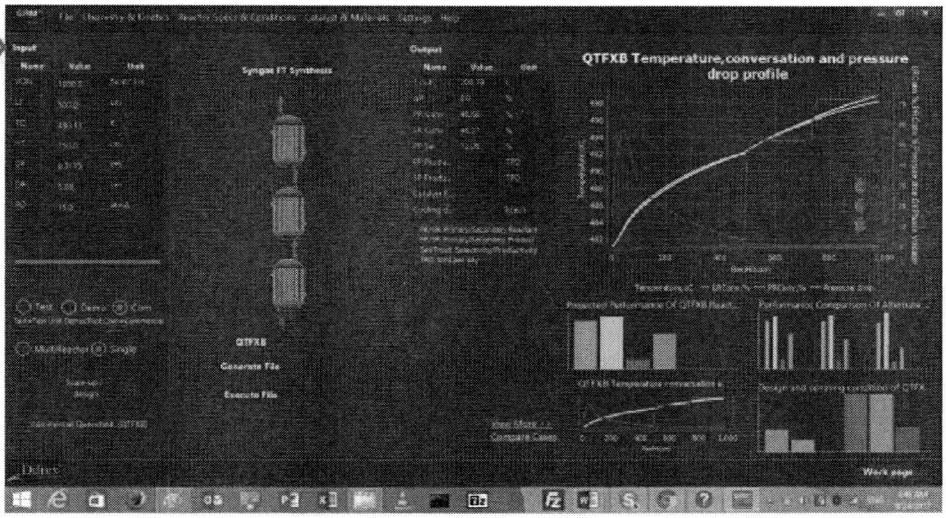

*Quick, friendly, easy and error-checked entries
@Complete temperature, pressure, composition, productivity, conversion and selectivity profiles, catalyst productivity and effectiveness factors, deactivation, equilibrium constants, reaction rates, cooling loads, thermo-chemical and transport property changes, heat and mass transfer modes, hydrodynamic behaviors like bubble growth, coalescence, slugging, grid, splash zone and free-board reactions, voidage profiles, TDH, solids loadings, flow regimes, entrainment rates, gas solid slip and more ....

Figure 16: Reactor design and development is as easy as 1-2-3 with DIFREX GRM™ software along with Reactor Of Choice selection

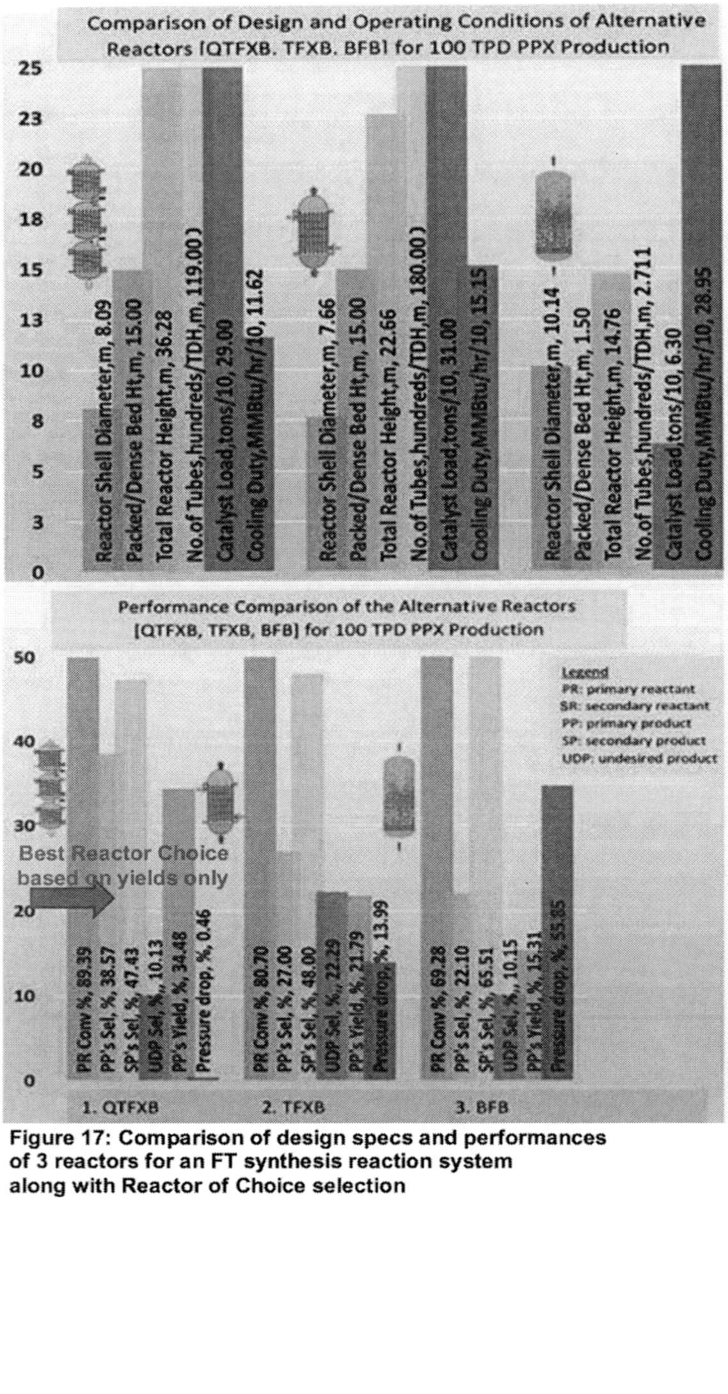

Figure 17: Comparison of design specs and performances of 3 reactors for an FT synthesis reaction system along with Reactor of Choice selection

Enterprise solution: Strategic/fast-track action and sustained growth

Fast-Track Technology: Helping to fast-track process and technology development is one of our core strengths. Our structured, industry-proven and 'no-oops' development strategy should get you to your goals faster. We did it before, Let us do it again … for you.

Disruptive change is the new norm that makes sustained growth challenging. Reactors in your plant face the brunt of this challenge as feeds, product demands and markets can change suddenly. DIFREX can help in your strategy to be proactive not only to minimize losses due to such disruptions, but also to capitalize on them.

DIFREX can also help infuse that vibrant **culture** into your strategy at both administrative and functional levels that deliver the strategic advantage – by fueling the spirits (desires, engagement, and execution) for all tasks around your reactors including R&D. We help you identify a **strategic niche** and make your strategic decision possible by our in-depth and unbiased analysis of both market potentials and true technical positions of competing/emerging technologies.

Integrated Technology Development (ITD): A New Concept

Traditionally, process design or technology development begins ***after*** determination of the Catalyst or Material Of Choice (COC/MOC) is reached. This determination is mostly made by seeking a stable (non- or slowly deactivating) catalyst or material that is most appropriate for a fixed-bed reactor only. For a fluidized-bed reactor, where a catalyst or material can be regenerated, such a constraint does not exist. A fast-deactivating, but otherwise advantageous catalyst or material, with high activity and selectivity, is perfectly suitable for a fluidized bed system if these can be regenerated and recycled. Therefore, it is advisable to proceed with determination of the Reactor Of Choice (ROC) or the process technology ***much sooner*** in the technology development path. This is *now* possible and can be accomplished easily. In the traditional approach, a reactor has to match the demand of the Catalyst or Material Of Choice, which often leads to the use of a fixed bed as the sole or first Reactor Of Choice. This could be a poor choice when compared with alternative reactors. Using the simultaneous or Integrated Technology Development (ITD) approach, the right reactor and/or the technology can be determined much faster.

5. Future of Business – Now

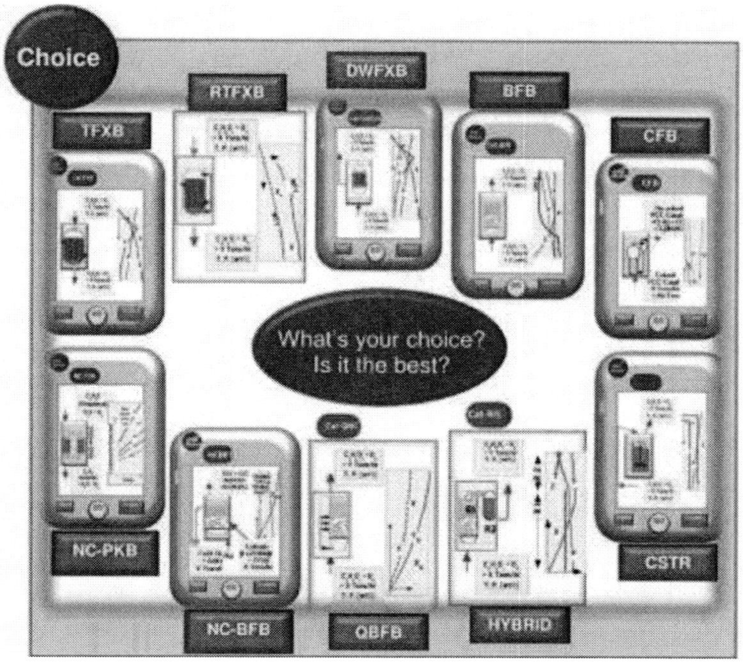

Figure 18: Alternative Designs and Development Paths now made easy

Process Technology Development will not be the same

It will be better, easier and much faster, leading to a quantum jump in *productivity* of design and development teams. Much of the task could be accomplished *quickly* in a hand-held device as shown schematically in **Figure 18** - with all security shields in place.

Micro-Channel Reactor (μ-CR) Technology [Process Intensification]:

For small to medium scale plants a micro-channel reactor (**Figure 19**) offers an advantageous alternative due to many attributes that include a) compactness, minimal footprint and size, b) negligible heat and mass transfer resistances, c) modularity, hence also load-following capability, d) shop fabrication, easy transportability and installation, e) minimum catalyst need and easy regenerability, and f) distributed applications. DIFREX provides services in building energy distributed networks for liquid fuel generation and supply, using microchannel reactors and renewable resources.

Figure 19: Micro-channel reactors for GTL applications

Value Chain Reactors: Fuels and chemicals from Syngas

Value chain reactors (**Figure 20**) can be easily and promptly developed and commercialized.

PTF (Power to Fuel) Technology

The cost of solar power, the most sustainable power source, is dropping and is contributing strongly to the US and world economy. According to one report, in 2016 the American solar industry added jobs 12x faster than the rest of the economy. There is a worldwide drive, particularly in Europe, to capitalize on the cheaper solar power to produce fuels and chemicals. Another area of key interest is to convert both powerplant exhaust and atmospheric CO_2 to fuel, especially to *methanol* that is dubbed as the *fuel of the future*. An economic *PTM* (Solar *Power To Methanol*) technology for conversion of atmospheric CO_2 to methanol would accomplish *zero-carbon footprint* in power consumption and utilization. We should heed the call for developing this emerging technology.

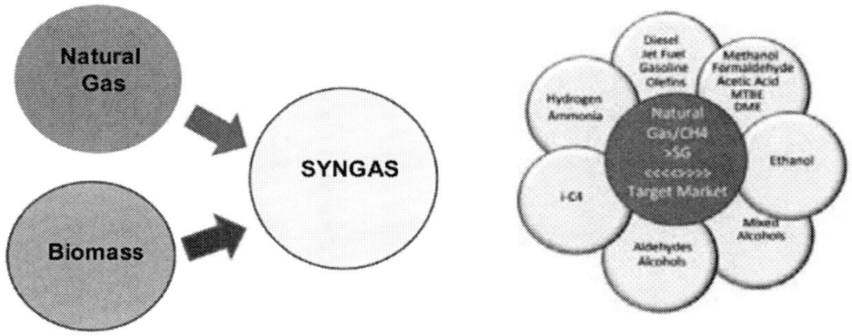

Figure 20: Conversion route of Value Chain Reactors

References:

1. Dutta, S and Gualy, R, "Build Robust Reactor Models", Chemical Engineering Progress, October, 2000, p. 37.
2. DIFREX[R] LLC NEXGEN Reactor Design - Fast assured reactor technology solutions for development, design, optimization, troubleshooting and revamp of commercial reactors. [cited November 13, 2017]; available from:www.difrex.com.
3. Dutta, S. and R. Gualy, "Super-modeling Reactors for Profitability", Chem. Eng'g. 107(6), 2000, p. 72.
4. Dutta, S. and R. Gualy, "Modernize Process Reactors", Hydrocarbon Processing, November, 1999, p. 91.
5. Dutta, S. and R. Gualy, "Overhaul Process Reactors", Hydrocarbon Processing, September, 1999, p. 43.
6. Dutta, S. and R. Gualy, "General Reactor Model Improves HPI Applications", Hydrocarbon Processing, July, 1999, p. 45.
7. Dutta, S., "Circulating Versus Bubbling Fluidized Bed for Hot Gas Desulfurization by Zinc Titanate", Proc. 5th Int'l. Circulating Fluidized Bed Conf., Beijing, PRC, May, 1996, p. pr-7.
8. Dutta, S., "Reactor Model: Usefulness of a Lesser Known Tool in Process Plants", paper presented at SABIC Tech. Meeting, May, 1996.
9. Dutta, S., "Fluidized and Moving Bed Desulfurization by Zinc Titanate", AIChE Symp. Ser. 301 (90), 1994, p. 157.
10. Dutta, S. and B. Jazayeri, "Alternate Reactor Concepts for Oxidative Coupling of Methane", Proc. 7th Int'l. Fluidization Conf., ed. Potter and Nicklin, Brisbane, Australia, May, 1992, p. 445.
11. Dutta, S. and G. D. Suciu., "An Experimental Investigation on the Effectiveness of Baffles in Breaking Bubbles in Fluid Beds", J. Chem. Eng'g. Japan 25 (3), 1992, p. 345.
12. Dutta, S., "A Novel CFB Application: A Chemical Compressor", Proc. 3rd Int'l. Circulating Fluidized Bed Conf., ed. Basu, Horio and Hasatani, Nagoya, Japan, Oct., 1990, p. 465.
13. Dutta, S., "A Novel Circulating Bed Metal Hydride Compressor", paper presented at the 8th World Hydrogen Conf., Honolulu, Hawaii, 1989.
14. Dutta, S., B. R. Christian and M. F. Ratterman, "Circulating Fluid Bed Reactor Model Developed for FCC/ART Catalyst Regenerator", Proc. 6th Int'l. Fluidization Conf., ed. Grace, Shemilt and Bergougnou, Banff, Canada, May, 1989, p. 9.
15. Dutta, S. and G. D. Suciu, "Unified Model Applied to the Scale-up of Catalytic Fluid Bed Reactors of Commercial Importance", Proc. 6th Int'l. Fluidization Conf., ed. Grace, Shemilt and Bergougnou, Banff, Canada, May, 1989, p. 311.
16. Dutta, S., S. C. Arnold, G. D. Suciu and L. Verde, "Scale-Up of a Commercial Catalytic Fluid Bed Reactor Involving Complex Kinetics", Ind. Chem. Eng'g. Symp. Ser. 87, 1984, p. 517.
17. Dutta, S., "Measurement of Bubble Characteristics by Optical Probe", paper presented at the AIChE Annual Meeting, Chicago, Nov., 1980.
18. Dutta, S., "A Simple Probe for Fluidized Bed Measurements", Cana. J. Chem. Eng'g. 57, 1979, p. 115.
19. Dutta, S., "Dead-zone Heights near the Grids of Fluidized Beds", Fluidization, Cambridge University Press, 1978, p. 32.
20. Wen, C. Y. and S. Dutta, "Research Needs for Analysis, Design and Scale-up of Fluidized Bed Reactors", AIChE Symp. Ser. 73 (161), 1977, p. 1.
21. Fluidization Engineering, 2nd edition, by Daizo Kunii and Octave Levenspiel, Butterworth-Heinemann, 1991.
22. Chemical Reaction Engineering, 3rd edition, by Octave Levenspiel, John Wiley & Sons, 1999.
23. Werther, J., Chem. Eng'g. Sci. 35, 1980, p. 372.
24. Kato, K., and C. Y. Wen, Chem. Eng'g. Sci. 24, 1969, p. 1351.
25. Hatano, H., T. Okuma and M. Ishida, J. Chem. Eng'g. Japan 17, 1984, p. 491.
26. Kwauk, M. N., N. Wang, Y. Li, B. Chen and Z. Shen, Circulating Fluidized Bed Technology, ed. P. Basu, Pergamon Press, 1986, p. 33.
27. Yang, W. C. and D. L. Keairns, AIChE Symp. Ser. 141(70), 1974, p. 27.
28. Wen, C.Y. and S. Dutta, Book Chapter "Kinetics of Coal Pyrolysis and Gasification Reactions", Ch. 2, Coal Conversion Technology, Addison-Wesley Publishing Company, 1979.
29. Wen, C.Y. and S. Dutta, "Solid-Gas Reactions in Coal Conversion Processes", "Coal Processing Technology", CEP Tech. Manual IV, 1978, p. 40.
30. Wen, C.Y. and S. Dutta, "Reactivity of Coal and Char: Part I – In Carbon Dioxide Atmosphere", I&EC Process Design & Development 16, 1977, p.20.
31. Wen, C.Y. and S. Dutta, "Reactivity of Coal and Char: Part II – In Oxygen-Nitrogen Atmosphere", I&EC Process Design & Development 16, 1977, p.31.
32. www.difrex.com.
33. http://difrex.com/pdf/BrochureContents.pdf
34. Dutta, Subhash, Stephen Arnold, Aashish Gaurav and Jim Brenner, "DIFREX Reactor and Technology Solutions for Many Particle Types and Sizes in Catalytic and Non-Catalytic Reaction Systems", paper presented at WCPT8 (8th World Congress on Particle Technology), Orlando, April, 2018.

Optimization of extraction condition for phytic acid from peanut meal by response surface methodology

*Hong Ren *, Ting Li , Huijie Wan, Jinping Yue*

Beijing Technology & Business University, School of Food and Chemical Engineering, Beijing Higher Institution Engineering Research Center of Food Additives and Ingredients, Beijing Key Laboratory of Flavor Chemistry, Beijing, China

Abstract Phytic acid (PA), a molecule with high commercial value, is one of the important component in peanut meal. However, PA has not yet been isolated from peanut meal to exploit and utilize. This paper reported the extraction conditions of PA from peanut meal after removed protein. The independent variables were hydrochloric acid (HCl) concentration, solid to liquid ratio, extraction time and extraction temperature. Response surface methodology (RSM) was used to optimize the extraction conditions based on the extraction yield of PA. The results show that the second-order polynomial models derived from responses well with the experimental ($R^2 = 0.9783$). The optimal extraction condition was obtained with solid to liquid ratio of 1:16 (g:mL), HCl concentration of 0.02 mol/L, extraction time of 105 min, and extraction temperature of 30°C. At this condition, PA with higher purity were obtained. the extraction ratio was 6.12%, and the content of PA was 182.7 mg/g dry PA extract. The experimental values under optimal condition were in good consistent with the predicted values. The PA extracted from peanut meal was verified qualitatively by IR spectra. The extraction technology of PA from peanut meal has a strong potential for realized high-value utilization of peanut meal.

Keywords: Phytic acid, Peanut meal, Response surface methodology, independent variables, optimization

1. Introduction

Myo-inositol hexakisphosphate, also called phytic acid (PA), is considered as an anti-nutrient component because of its strong ability to combine with multi-charged metal ions, specially Zn^{2+}, Ca^{2+} and Fe^{3+} [1,2]. Recently, a variety of pharmacological activity of PA have been reported. S. Norazalina found the potential value of PA extracted from rice bran in reducing colon cancer risk in rats [3]. The finding showed treatment with 0.2% (w/v) PA extract gave the greatest reduction in the formation of aberrant crypt foci. Yukako Okazaki's study showed that PA may improve the composition of cecal organic acids, microflora, and mucins, and it may decrease the levels of serum proinflammatory cytokines in rats fed a high-fat, mineral-sufficient diet[4]. PA is a potential absorption enhancer of flavonoid components on tight junction integrity in Caco-2 cell monolayers[5]. PA is a naturally occurring constituent which exhibits protective action in Parkinson's disease and it has been shown to lower blood glucose levels[6,7]. Moreover, PA has been widely used in metallic material field as a new generation of green corrosion inhibitor for copper [8-11]. cupronickel B30 [12], brass [13] and Mg-Li alloys[14]. PA is widely used in food and light industry.

PA mainly distributes in seeds of plants, such as cereal grains, legumes, nuts, oilseeds, and so on, acting as a main source of phosphorus. Many cereal grains and oilseeds contain about 1~3% PA [15]. Peanuts is one of the most widely grown oil seeds in the world with 29 million tons produced every year [16]. In China peanuts are the most important oil seeds, ranking first among the world's peanuts producing countries[17]. Nowadays, a majority of peanuts are crushed to produce oil and this process generates a large amount of peanut meal as the by-product, which is about 9 million tons every year in china [18-20]. A large number of studies showed that peanut meal is a good source of active ingredients, which contains about 47~55% protein, 20~30% carbohydrates, 8~10% crude fiber, 2~3% fat, and 1.0~1.2% PA [21]. However, most of the available peanut meals are used as feed ingredient for animals, and only a small portion was used to recover proteins, causing great waste [22]. Therefore, obtaining the active ingredients and high-value utilization of peanut meal has become increasingly important task for effectively reducing peanut resources waste.

Compared with other plant materials, peanut meal is rich in PA, but there are no good process to isolate PA from peanut meal reported so far. The effective extraction procedure and method to separate PA from peanut meal are very necessary. In this study, we firstly removed proteins from peanut meal, and use response surface methodology (RSM) to optimize the extraction conditions. The objective of present study was to find the optimum conditions for the extraction of PA, including hydrochloric acid concentration, extraction temperature, extraction time and ratio of liquid to material, to maximize the extraction rate of PA from the peanut meal after removing proteins.

2. Materials and methods

2.1 Materials

Peanut meal was provided by Qinglonghu Wanchunyuan Vegetable Center in Beijing. The peanut meal was dried at 50°C in an oven after removing water-soluble proteins. The processing steps of removing proteins from peanut meal is shown in **Fig.1**. The dried peanut meal after removing proteins was ground using a mill.

Phytic acid sodium salt hydrate was purchased from Sigma-Aldrich Co. (St. Louis, MO, USA). Ferric trichloride was purchased from Tianjin Samtec Chemical Reagent Co., LTD. 5-Sulfosalicylic acid dehydrate was purchased from West Gansu Chemical Plant in Shantou city. HCl was purchased from China National Pharmaceutical Group Chemical Reagent Co., LTD.

2.2 Determination of PA content

The determination of PA content was evaluated by the Ferric trichloride combined 5-Sulfosalicylic acid method according to the procedure described previously[22]. 0.15g Ferric trichloride and 1.5 g 5-Sulfosalicylic acid was dissolved in 500 mL waters to be the Ferric trichloride and 5-Sulfosalicylic acid reaction solution. Inhale 0.0, 1.0, 2.0, 3.0, 4.0, 5.0 mL of 0.1 mol/L Phytic acid sodium salt hydrate solution in 6 colorimetric cylinders respectively, and trickle water into the colorimetric cylinders to 5 mL. A 5 mL Phytic acid sodium salt hydrate solution was combined with 4 mL Ferric trichloride and 5-Sulfosalicylic acid reaction solution. The absorbance of the solution was measured using a UV-vis spectrophotometer (754 UV-vis Spectrophotometer, Shanghai Jinghua Co, LTD) at a wavelength of 500 nm. Standard curve was draw with the absorbance as the ordinate and PA content as the ordinate. The regression equation was obtained as the following equation:

$$Y= -4.126x+0.878 \quad (R^2=0.9991) \qquad (1)$$

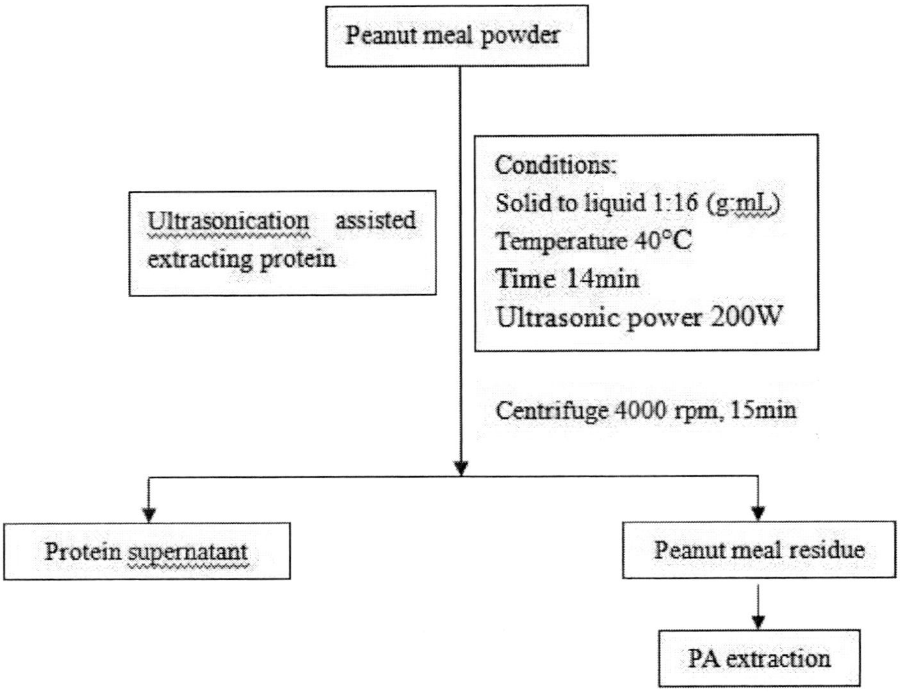

Fig. 1 The protein removal steps from peanut meal

2.3 Preparation of solid-liquid extracts

The influence of the solid-to-liquid ratio on the extraction of PA was investigated by using the following six ratios (1:10, 1:12, 1:14, 1:16, 1:18, 1:20; g:mL; sample powder: solvent). The hydrochloric acid concentration was fixed at the concentration of 0.01mol/L. The mixtures were extracted at 32°C for 83 min. The extraction solution was centrifuged at 4000 rpm for 10 min to obtain the supernatant. The content of PA (mg PA/g dry peanut meal) was then determined. The ratio that gave the highest value of yield was chosen for RSM.

2.4 Experimental designs

A d-optimal RSM experiment was designed using Design Expert (8.0.6, Stat-Ease Inc., USA). The independent variables were hydrochloric acid concentration (mol/L, X_1: 0-0.04), extraction time (min, X_2: 20-140), extraction temperature (°C, X_3: 25-60). The operating conditions were selected after single factor experiment and shown at **Table1**. The response variable was fitted by a second-order polynomial as follows:

$$Y = \beta_0 + \sum_{i=1}^{3} \beta_i X_i + \sum_{i=1}^{3} \beta_{ii} X_i^{\ 2} + \sum_{i=0}^{2} \sum_{j=i+2}^{3} \beta_{ij} X_i X_j \qquad (2)$$

Y : the predicted response; β_0: the intercept coefficient; β_i: the linear coefficient; β_{ii}: the squared coefficient; β_{ij}: the interaction coefficient; X_i, X_j: the coded independent variables; X_iX_j: the interaction terms; X_i^2: the quadratic terms.

The optimal conditions for the extraction of PA from peanut meal were then carried out using the equations of RSM.

Table 1 Independent variables and their levels used for Box-Behnken rotatable design

Independent variable	**Level**		
	-1	**0**	**1**
HCl concentration (X_1)	0.01	0.02	0.03
Time (X_2)	80	100	120
Temperature (X_3)	25	30	40

2.5 The extraction rate and purity of PA

The extraction rate of PA is the ratio of the weight of PA crude extract to the weight of peanut meal, and the purity of PA is the ratio of the pure PA to the PA crude extract. The calculation formulas of the extraction rate and purity are as follows:

The extraction rate (%) = (The weight of PA extract / The weight of peanut meal) × 100% (3)

The PA purity (%) = (The weight of PA / The weight of PA extract) × 100%   (4)

2.6 Fourier Infrared spectrum analysis the PA of peanut meal

FT-IR measurement was recorded by a AVATAR 370 (Thermo Nicolet). 1mg dried PA sample was mixed with 200 mg KBr. The mixture was milled 5 min in the agate mortar and tablet. The spectra of each sample were used to analysis in the range of 400-4000 cm^{-1}.

3. Results and discussion

3.1 Determination of single factor experiment

3.1.1 Effect of solid/liquid ratio

In order to optimize the extraction of PA from peanut meal, solid to liquid had to be selected to study [23]. The ratio of solid to liquid varied from 1:5 to 1:20 (g:mL) while other extraction variables were fixed. In **Fig.2a**, as the decrease of the solid to liquid ratio, the content of PA increased. The highest yield of PA was found at 1:16 (g:mL). This indicated that more volume of solvent was required to sufficiently solubilize the target PA when the higher driving force of molecules was supposed to be higher in a high solid to liquid ratio. However, there was no significant differences in extraction yield from 1:16 to 1:20 (g:mL) ($p > 0.05$). Therefore, 1:16 (g:mL) was chosen as a compromise and fixed in the RSM experiment.

3.1.2 Effect of HCl concentration

In acid solutions, the complexing power of PA is declined, and the solubility of PA is increased. The effect of HCl concentration on the content of PA is shown in **Fig.2b**. The highest extraction yield of PA was observed at HCl concentration of 0.02 mol/L. From 0 to 0.02 mol/L, the content of PA increased. While from 0.02 to 0.04 mol/L, the content of PA decreased. As increased of HCl concentration, the PA was dissolved in the HCl solution because of the decline of complexing capacity of PA. When the HCl concentration was beyond the optimized concentration, other acid soluble components, such as acid soluble saccharide, amino acids, could be dissolved, may recombine with PA and form complexes, resulting in the decrease of the content of PA.

3.1.3 Effect of extraction time

The effect of extraction time on the content of PA was studied varying time from 20 to 140 min. The content of PA was substantially increased with the time rose from 20 to 100 min (**Fig.2c**). It had a decrease at longer time. It could be that other acid soluble components were dissolved and recombined with PA. The optimized extraction time was 100 min.

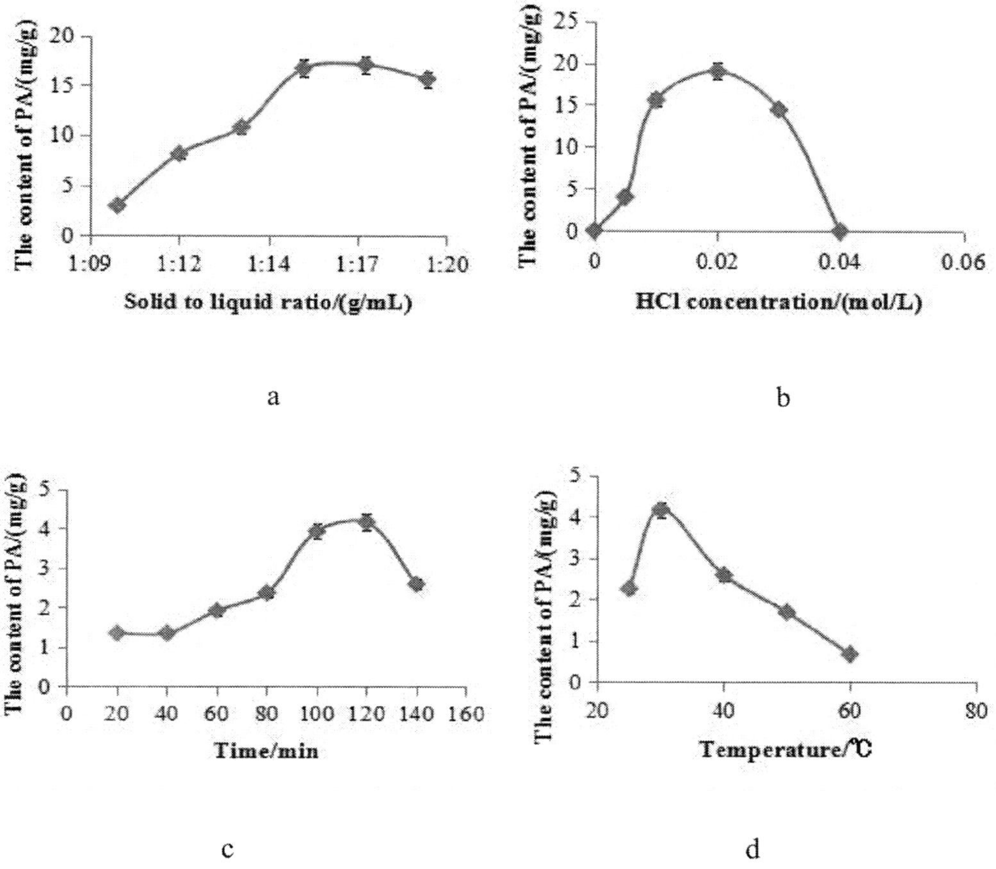

Fig.2 Single factor experiment

Effect of solid to liquid ratio (a), HCl concentration (b), extraction time (c) and Temperature (d) of PA from peanut meal. Different letters in the same sample indicate significant differences ($p < 0.05$) between treatments.

3.1.4 Effect of extraction temperature

As shown in **Fig.2d**, the yield of PA was increased with increased of extraction temperature from 25 °C to 30 °C. The maximum value of extraction temperature was at 30°C, having an obviously decrease with a further increase of extraction temperature. As the temperature increased, other ingredients were dissolved and combined with PA, causing the decreased of extraction yield.

According to the findings of above studies, the HCl concentration of 0.02 mol/L, extraction time of 100 min, extraction temperature of 30°C were selected as a central point for RSM. The ratio of solid to liquid was fixed at 1:16 (g:mL) in the following experiment.

Table 2 Coded levels, conditions runs and measured responses used in experimental design for response surface methodology

Run	X_1	X_2	X_3	**Yield (mg PA/g peanut meal)**
1	0.02	100.00	32.50	13.48
2	0.02	80.00	40.00	9.30
3	0.02	120.00	40.00	10.27
4	0.03	100.00	25.00	1.85
5	0.02	100.00	32.50	13.36
6	0.01	100.00	40.00	4.26
7	0.02	100.00	32.50	11.92
8	0.02	80.00	25.00	9.09
9	0.03	120.00	32.50	0.90
10	0.02	100.00	32.50	12.59
11	0.01	100.00	25.00	9.87
12	0.01	120.00	32.50	8.52
13	0.03	100.00	40.00	6.45
14	0.03	80.00	32.50	1.80
15	0.02	120.00	25.00	10.56
16	0.01	80.00	32.50	5.87
17	0.02	100.00	32.50	13.22

Table 3 Analysis of variance (ANOVA) for the fitted quadratic polynomial model for optimization of extraction parameters

Source	Sum of Square	Degree of freedom	Mean square	F value	p-value p>F	Significance [a]
Model	279.20	9	31.02	35.12	<0.0001	**
X_1	38.36	1	38.36	43.43	0.0003	**
X_2	2.20	1	2.20	2.49	0.1586	
X_3	0.15	1	0.15	0.17	0.6930	
X_1X_2	3.15	1	3.15	3.56	0.1010	
X_1X_3	26.05	1	26.05	29.50	0.0010	**
X_2X_3	0.062	1	0.062	0.071	0.7982	
X_1^2	173.67	1	173.67	196.60	<0.0001	**
X_2^2	20.78	1	20.78	23.52	0.0019	**
X_3^2	3.30	1	3.30	3.74	0.0945	
Residual	6.18	7	0.88			
Lack of Fit	4.47	3	1.49	3.47	0.1301	Not Significant
Pure Error	1.72	4	0.43			
Cor Total	285.38	16				

R^2=0.9783, Adj R^2=0.9505, Pred R^2=0.7401, adeq precision=16.526.

[a] *significant ($p < 0.05$), **extremely significant ($p < 0.01$)

3.2 Response surface analysis

3.2.1 Analysis of the adequacy of the fitted model

RSM was applied to determine the effect of the HCl concentration (X_1), extraction time (X_2), and temperature (X_3) on the content of PA (Y). The responses of each independent variable are listed in **Table 2**. The mathematical model for describing the content of PA was given by the following second-order polynomial equation:

$$Y=12.91-2.19X_1+0.52X_2-0.14X_3-0.89X_1X_2+2.55X_1X_3-0.12X_2X_3-6.42X_1^2-2.22X_2^2-0.89X_3^2$$

The response surface quadratic model was drawn up, and the analysis of variance for the fitted quadratic polynomial model for optimization of extraction parameters are presented in Table 3. The p-value for the model was less than 0.0001, which indicated the model was extremely significant and could be used to monitor the optimization. X_1, X_1X_3, and quadratic terms (X_1^2 and X_2^2) were extremely significant on the PA extraction value.

As **Table 3** shown that the "Lack of Fit" is not significant relative to the pure error. The coefficient of multiple determinations (R^2) for the quadratic regression model was 0.9783. The R^2 was higher than 0.7, indicated that the model was suitable for use in the experiment [24]. The value of adjusted determination coefficients (R^2_{adj}) was 0.9505, closed to the R^2. Therefore, this model can be used to navigate the design space [25].

3.2.2 Response surface and contour plots

The response surface plots and contour plots shown in Fig.3. It demonstrated the effect and interaction of independent variables on the yield of PA. Fig.3 a-1 and a-2 shows the effect of HCl concentration and extraction time on the content of PA. when the extraction time was fixed, the content of PA gradually increased with the increase of HCl concentration, up to the HCl concentration of 0.02 mol/L , a large increment of PA content was observed. This due to the increase of HCl concentration improved the solubility of the acid soluble components. When the HCl concentration was fixed, an obvious quadratic effect of extraction time was observed. Fig. 3 b-1, b-2, and Table 3 demonstrates the interaction of HCl concentration and extraction temperature is significant ($p < 0.05$). The content of PA as affect by HCl concentration and extraction temperature, exhibiting a clear increase in content with the raise of HCl concentration at a fixed extraction temperature. While an obvious quadratic effect of extraction temperature was observed. In **Fig.3** c-1 and c-2, a similar linear increase in content with the increase of extraction time ranged from 80 min to 104 min at a fixed temperature. While a decrease in temperature from 104 min to 120 min led to marked decrease in content of PA. This indicated that the extraction yield of PA depended on the medium level of time. These occurrences might be due to the recombination of PA and acid soluble proteins.

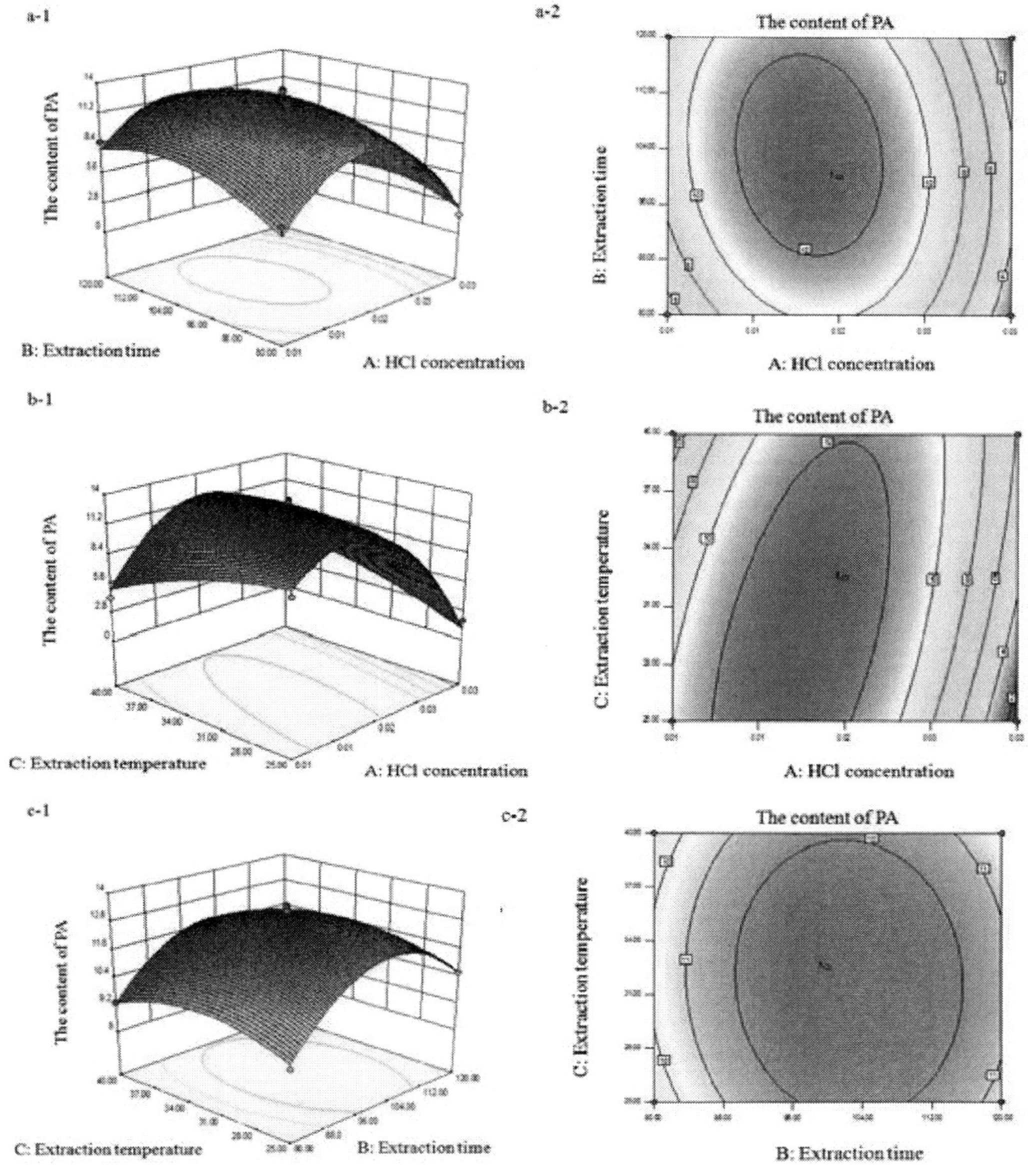

Fig. 3 Response surface plot and contour plot showing the effects of independent variables on the content of PA. a. HCl concentration and extraction time, b. HCl concentration and extraction temperature, c. extraction time and extraction temperature

3.2.3 Optimization of extraction craft parameters and the model validation

The optimal extraction conditions for PA were determined in order to simultaneously maximize the extraction. It was found that the optimal conditions were an HCl concentration of 0.02 mol/L, extraction time of 105 min, extraction temperature of 30 °C. The predicted and experimental values for the extraction of PA from peanut meal were 13.31 mg/g and 11.18 mg/g, respectively. In the optimal conditions, the experimental value was close to the predicted value, and the extraction ratio and content of PA was 6.12 ± 0.51% and 182.70 ± 2.35 mg/g, respectively. Removing the protein firstly and using the optimum extraction conditions can improve the extraction yield and purity. These results confirmed the predictability of the model to give the maximum yield of PA in the experimental process used.

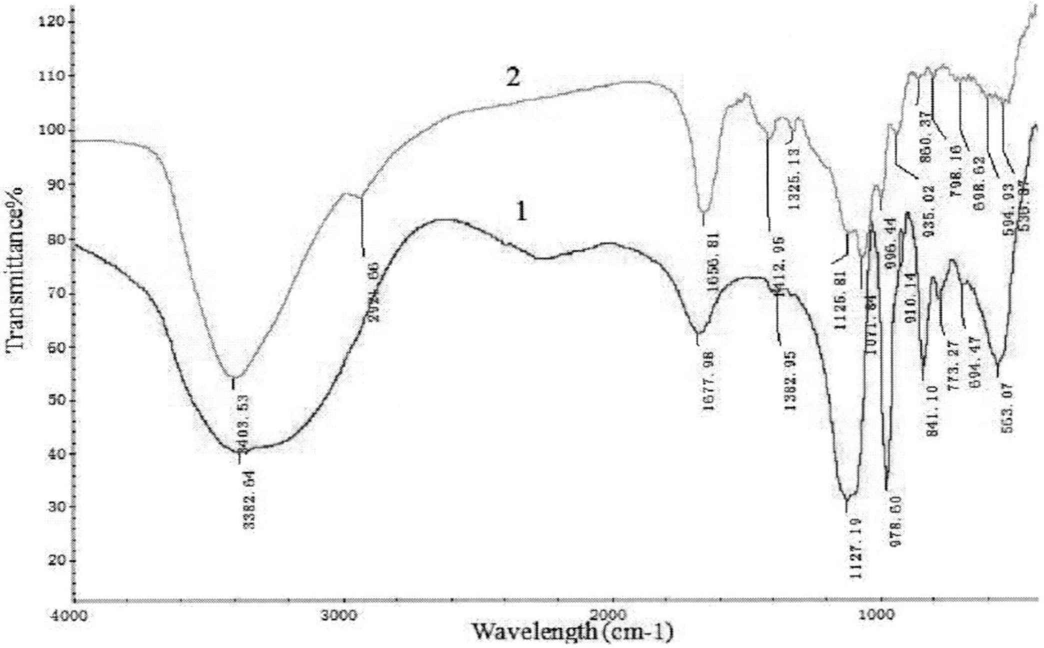

Fig.4 FT-IR spectra of PA extract.

1: Reference material phytic acid sodium salt hydrate from Sigma; 2: PA extract from peanut meal

3.3 Fourier Infrared spectrum analysis the PA of peanut meal

Each FT-IR spectra site of a compound provides qualitative information. PA contains phosphoester bond (P=O, P-O, C-O, the region of 1330~900 cm^{-1}; -OH, the region of 3650~3200 cm^{-1}). **Fig.4** was shown that reference material phytic acid sodium salt hydrate and PA extract from peanut meal have same peaks. Phytic acid sodium salt hydrate has the peaks at the wavelength of 3382 cm^{-1} (-OH), 1127 cm^{-1}, 978 cm^{-1} (P=O, P-O, C-O). PA extract has peaks at the wavelength of 3403 cm^{-1} (-OH), 2921 cm^{-1} (-CH), 1125 cm^{-1}, 1071 cm^{-1}, 996 cm^{-1} (P=O, P-O, C-O). Therefore, the PA extract from peanut meal was verified qualitatively by comparing with phytic acid sodium salt hydrate in the IR spectra.

4 Conclusion

In this study, we first removed proteins from peanut meal, and RSM was used to optimize the experimental variables including HCl concentration (mol/L), extraction time (min), and extraction temperature (°C). Solid to liquid ratio had significant effect on the extraction yield at the ratio of 1:16 (g:mL). This was chosen as the optimum parameter. The optimal conditions to obtain the highest extraction of PA were determined as follows: 0.02 mol/L of HCl concentration, 105 min of extraction time, 30°C of extraction temperature. At this condition, the extraction ratio was 6.12 ± 0.51%, which is far higher than the extraction ratio of traditional method, and the content of PA was 182.70 ± 2.35 mg/g dry PA extract. The PA extract derived from peanut meal could be verified qualitatively by IR spectra.

PA is the most important anti-nutrient ingredient in peanut meal. Peanut nutrient ingredients removing PA is one of the most attractive and promising sources of plant proteins or plant polysaccharides. On the other hand, peanut meal is the important renewable source of PA, PA extracted from peanut meal will be better applied in medicine and light industry. This study could not only be useful in science research and industry production of PA from peanut meal, but also conducive to high value utilization of peanut meal and reduce waste of peanut resources.

Acknowledgements

This work was supported by the National Key Research and Development Program of China (2016YFD0400802), National high technology research and development program of China (863 Program) (2011AA060701).

References

[1] Wang H., Zhou Y., Ma J., Zhou Y Y., Jiang H (2013). The effects of phytic acid on the Maillard reaction and the formation of acrylamide. Food Chemistry, 141 (1) , 18-22.

[2] Lopez H.W., Leenhardt F., Coudray C., Remesy C. (2002). Minerals and phytic acid interactions: is it a real problem for human nutrition? Int. J. Food Sci. Technol., 37, 727-739.

[3] Norazalina S., Norhaizan M.E., Hairuszah I., Norashareena M. S. (2010). Anticarcinogenic effcacy of phytic acid extracted from rice bran on Azoxymethane-induced colon carcinogenesis in rats. J. Experimental and Toxicologic Pathology, 62 (3), 259-268.

[4] Yukako O., Tetsuyuki K. (2014). Dietary phytic acid modulates characteristics of the colonic luminal environment and reduces serum levels of proinflammatory cytokines in rats fed a high-fat diet. Nutrition Research, 34 (12) , 1085-1091.

[5] Fu Q. X., Wang H. Z., Xia M. X. (2015). The effect of phytic acid on tight junctions in the human intestinal Caco-2 cell line and its mechanism. J. European Journal of Pharmaceutical Sciences, 80, 1-8.

[6] Lv Y. Q., Zhang Z., Hou L., Zhang L., Zhang J. Y., Wang Y. H., Liu C., XuP . P., Liu L., Gai X. Y., Lu T. X. (2015). Phytic acid attenuates inflammatory responses and the levels of NF-kB and p-ERK in MPTP-induced Parkinson's disease model of mice. J. Neuroscience Letters, 597, 132-136.

[7] Jin N. K., SungN. H., Kim H. K. (2014). Phytic acid and Myo-inositol support adipocyte differentiation and improve insulin sensitivity in 3T3-L1 cells. Nutrition Research, 34, 723-731.

[8] Notoya T., Otieno-Alego V., Schweinsberg D.P. (1995). The corrosion and polarization behaviour of copper in domestic water in the presence of Ca Mg and Na-salts ofphytic acid, Corros. Sci. 37 (1), 55-65.

[9] Yang H.F., Yang Y., Yang Y. H.(2005). Formation of inositol hexaphosphate monolayers at the copper surface from a Na-salt of phytic acid solution studied by in situ surface enhanced Raman scatteringspectroscopy Raman mapping and

polarization measurement. Anal. Chim.Acta, 548 (1-2), 159-165.

[10] Wang Y. .H., He J. B. (2012). Corrosion inhibition of copper by sodium phytate in NaOH solution: cyclic volt absorptometry for in situ monitoring of soluble corrosion products. Electrochim. Acta. 66, 45-51.

[11] Li C. C., Guo X.Y., Shen S. (2014). Adsorption and corrosion inhibition of phytic acid calcium on the copper surface in 3 wt% NaCl solution, Corros. Sci., 83, 147-154.

[12] Hao C., Yin R. H., Wan Z. Y., Xu Q J., Zhou G. D. (2008). Electrochemical and photoelectrochemical study of the self-assembled monolayer phytic acid on cupronickel B30, Corros. Sci., 50 (12), 3527-3533.

[13] Li Y., He J. B., Zhang M. (2013). Corrosion inhibition effect of sodium phytate on brass in NaOH media. Potential-resolved formation of soluble corrosion products. Corros. Sci., 74, 116-122.

[14] Gao L. L., Zhang C. H., Zhang M. L. (2006). Phytic acid conversion coating on Mg-Li alloy. Journal of Alloys and Compounds, 485 (1-2) (2009), 789-793.

[15] H. R. Park, H. J.Ahn, S. H.Kim, C. H.Lee, M. W. Byun, G. W. Lee. Determination of the phytic acid levels in infant foods using different analytical methods. Food Control, 17 (9), 727-732.

[16] Nader J., Fawaz N., Afif C., Louka N. (2016). A novel process for preparing low-fat peanuts: Optimization of the oil extraction yield with limited structural and organoleptic damage. Food Chemistry, 197, 1215-1225.

[17] Su G. W., Ren J. Y., Yang B. (2011). Comparison of hydrolysis characteristics on defatted peanut meal proteins between a protease extract from Aspergillus oryzae and commercial proteases. Food Chemistry, 126 (3) , 1306-1311.

[18] Wang Q., Shi A., Liu H., Liu L., Zhang Y., Li N., Gong K., Yu M., Zheng L. (2016). Chapter 5-Peanut By-Products Utilization Technology. Peanut: Processing Technology and Product Development, Copyright © 2016 China Science Publishing & Media Ltd. Published by Elsevier Inc., 211-325.

[19] Reddy N., Chen L.H., Yang Y. Q. (2013). Thermoplastic films from peanut proteins extracted from peanut meal. Ind. Crop. Prod. 43, 159–164.

[20] Wu H. W., Wang Q., Ma T. Z. (2009). Comparative studies on the functional properties of various protein concentrate preparations of peanut protein. Food Research International, 42 (3), 343-348.

[21] Zheng L., Ren J. Y., Su G. W., Yang B., Zhao M. M. (2013). Comparison of *in vitro* digestion characteristics and antioxidant activity of hot- and cold-pressed peanut meals. Food Chemistry, 141 (4), 4246-4252.

[22] Kiassos E., Mylonaki S., Makris D. P., Kefalas P. (2009). Implementation of response surface methodology to optimise extraction of onion (Allium cepa) solid waste phenolics. Innovative Food Science and Emerging Technologies, 10 (2), 246-252.

[23] Sahin S., Samli R. (2013). Optimization of olive leaf extract obtained by ultrasound-assisted extraction with response surface methodology. Ultrason Sonochem, 20 (1), 595-602.

[24] Wong K. H., Li G. Q., Li K. M., Valentina R. N., Chan K. (2017). Optimisation of Pueraria isoflavonoids by response surface methodology using ultrasonic-assisted extraction. Food Chemistry, 231 (15), 231-237.

[25] Bancha Y., Nuttapum S. (2015). Optimization of process parameters for phenolics extraction of Cratoxylum formosum ssp. formosum leaves by response surface methodology. Food Science Technology, 52 (1), 129-140.

512826 Influences of Geometric Factors on a Solar Thermochemical Reactor for Two-Step CO_2 Splitting

Abstract:

Solar thermochemical processes utilize the inexhaustible solar energy as thermal driving force to provide heat and motivate the reactions, which offer pathways to store solar energy to chemical fuels. Some elevated temperature reactions, like synthesis of syngas and methane reforming, are anticipating to accomplish in solar reactors, which have the ability to provide extremely high temperatures. Presently, the dominating factors in the design of solar reactors are heat and mass transfer with reaction mechanisms. A novel partition cavity-receiver reactor concept is proposed in this paper. For the purpose to provide a longer pathway of interaction between catalyst and reactants, a partition is introduced in this cavity-receiver reactor. A numerical computational fluid dynamics (CFD) analysis is performed to study on the influences of geometric factors, like length-diameter ratio, partition length and thickness, the transparency of partition, to fluid flow and heat transfer under constant solar power input for both two-dimensional and three-dimensional models. A two-step solar thermochemical redox reaction using ceria as catalyst to split CO_2 is modeled in the partition cavity-receiver reactor to investigate the relationship between geometric factors and reaction rates. Based on the comparisons and analysis of results, optimized geometric factors and corresponding operating conditions are discussed.

512577 Modular Systems for Clean Coal

Abstract:

There Are Many Technical and Manufacturing Challenges That Face Our Collective Efforts to Build an Affordable, Clean, and Secure Energy Future. To This End, the U.S. Department of Energy (USDOE) Has Funded Research and Development of Technologies Focused on Assessing Strategies for Rapid Deployment of Modular Systems That Will Expand Market Penetration of Clean Energy Technologies at All Scales. Â Currently Usdoe Is Assessing the Potential for Modular Systems to Accelerate Technology Deployment By Leveraging Advance Manufacturing Techniques That Make Such Systems Competitive. This Report Will Specifically Aim to Discuss Interchangeable, Interoperable, and Robust Modular System Platforms That Separate Gases (CO_2, N_2, O_2), Convert Domestic Carbon Resources (coal, natural gas, biomass, and biogas) to Produce Value-Added Products (power, fuels, and chemicals) While Addressing Environmental Concerns (water, carbon, and other emissions). Specifically the Report Will Identify the Major Challenges Facing Modularization and Examine Possible Technical and Economic Solutions through an Integrated Systems Perspective. the Report Also Discusses Opportunities to Harness Innovations in Advanced Manufacturing (e.g. additive manufacturing, microfluidics, advanced simulation) to Reduce Costs, Improve Performance, and Consider Design Elements to Improve Deployment. The Author Will Discuss the Challenges for Making These Modular Systems Come to Marketplace

517173 NSF Funding Opportunities for Fundamental Research on Modular Chemical Process Systems

Abstract:

This presentation will focus on NSFâ€™s vision and funding opportunities for fundamental research related to Modular Chemical Process Systems (MCPS). MCPS have emerged as a potentially transformative concept, powered by the implementation of principles of Process Intensification (PI), that can yield substantial economic and environmental benefits for certain applications, when compared to traditional chemical plants. For example, MCPS can enable rapid deployment of new chemical and energy technologies, take advantage of geographically scattered feedstocks, and provide autonomous operation. Additional potential benefits of MCPS include reduced capital cost, improved handling of variations in feedstock supply rate and composition, reduced plant down-time, higher energy efficiency, increased safety, and waste minimization. Successful implementation of this new chemical manufacturing paradigm at commercial scales depends on the development of novel conceptual designs based on PI, a fundamental understanding of the underlying complex chemical and physical processes taking place in intensified process units, the development of methods for cost-effective manufacturing of modules, and the development of algorithms for real-time optimization and control of MCPS. This is an opportunity for convergent fundamental research using principles from process systems engineering, catalysis, reaction engineering, molecular thermodynamics, separations, transport phenomena, data science, and smart/additive manufacturing.

517213 Chemical and Thermal Process Intensification at ARPA-E

Abstract:

The US Department of Energyâ€™s Advanced Research Projectâ€™s Agency â€“ Energy (ARPA-E) has been active in funding projects focused on the intensification of chemical and thermal processes for improvements in industrial energy efficiency. In this talk Fellow and Acting Program Director Addison K Stark will highlight major areas of ARPA-E investment and discuss ongoing interests in process intensification.

512543 Role of Qualitative Risk Assessments in Managing Project Start-Ups

Abstract:

A qualitative approach to performing risk assessments for project start-ups has recently been developed at Oak Ridge National Laboratory with the goal of providing a simple tool to help manage projects that require cross-functional teams to implement a project with a variety of drivers and/or constraints. This approach focuses on the development of risk-based planning to guide selection of tasks to be performed during the planning and start-up phases of a project to reduce risk, lower costs, and improve chances of success. Examples of how the tool has been successfully implemented will be given.

511697 Commissioning Planning - the Key to a Successful Project Startup

Abstract:

In most industrial capital investment projects, the final stage of manufacturing technology delivery involves some combination of factory acceptance test, site acceptance test and post installation support. The intent of these services is to allow buyers of manufacturing systems to assure that systems meet specifications and lack any performance defects. Performance, as defined in nearly all specifications, is based upon nominal conditions. Unfortunately, variation thrives in factories. The combination of raw material variance, human error, environmental factors, infrastructure fluctuations and process interdependencies can cause significant performance issues within newly installed manufacturing systems. Successful projects account for these natural fluctuations in processes and establish more than equipment acceptance tests, they incorporate a commissioning plan. The commissioning plan recognizes the importance of running new systems in their natural environment, complete with variation. Deliberate testing beyond process boundary conditions allow startup teams to establish a set of process controls that ensure high quality production at required throughput rates. This presentation will offer a template for developing successful commissioning plans and provide a few examples of industrial operations that leveraged effective commissioning.

513362 3 Superpowers to Boost Your Career

Abstract:

Career growth has little to do with how smart you are or how hard you work. Despite the long hours, we stay in our role, wondering when we will get promoted or when our efforts will be recognized. Often we believe that others are holding us back, preventing us from getting that promotion or recognition and causing us stress. The truth is that, beyond a performance threshold, success has more to do with ourselves holding us back. The good news about this is that we can control ourselves. In this workshop, I will share three key concepts and techniques that I have learned that have helped me move from a worker to a leader - completely changing my career trajectory and limitations as a result.

Engineers are often hired for their knowledge or technical skills. But over the course of a career, they need to develop professionally - developing new skills and changing others' perception in order to gain visibility and to be given the opportunities to expand professionally. The path to this career success can be distilled down to 3 core behaviors, which I call superpowers. They are not obvious on the surface, but they make a huge difference when put into practice. First, change your relationship with fear. By changing our relationship with fear we change the decisions that we make. Second, change your story. This changes the perception of your ambitions and contributions. Third, learn to breathe. The ability to take a breath and then act in a considered way helps you deal with the challenges of change and it also builds others' confidence in you.

Since beginning to employ these strategies, my career has taken off. I made significant contributions outside of my role. Over the next few years I was granted an $85,000 scholarship from my company to pay for an Executive MBA, at age 30 I was named a Leader Under 40 by Control Engineering magazine, and I was promoted 3 times in 4 years up to a Global Director role at age 33.

506005 "We Have That Covered": The Ten Most Common Mistakes in Pilot Plant and Laboratory Hazard Analysis

Abstract:

"We Have that Covered":

The Ten Most Common Mistakes in Pilot Plant and Laboratory Hazard Analysis

Richard Palluzi PE, CSP

Richard P Palluzi LLC

72 Summit Drive, Basking Ridge, NJ 07920

rpalluzi@verizon.net, 908-285-3782

Effective hazard analysis and risk assessment is always difficult in any situation. It is especially difficult in a research environment such as for pilot plants and laboratory operations. This paper will present the top ten problems the author has encountered in his 40 years of experience in this area. These include:

1. Failure to develop detailed credible scenarios
2. Not accepting the full potential for human error
3. Failure to thoroughly evaluate systems and controls as fail safe
4. Considering dependent failures as independent
5. Incomplete information on all the materials involved and their properties
6. Tacitly assuming the proposed work is a trivial risk
7. Approaching the review in too unstructured a manner
8. Assuming that more reviewers equates to a better review
9. Not recognizing something as a change
10. Willingness to compromise on unworkable solutions

The presentation will highlight each area and suggest some potential corrective actions.

510448 New Software Tools for Managing Change in a Laboratory Environment

Abstract:

The Management of Change (MOC) process for the Research & Technology organization of Chevron Phillips Chemical Co. LP was overhauled to strategically align all research facilities under one MOC umbrella. The process was organized into five key areas that focus on Personnel, Chemicals, Laboratory Hazards, Digital Control Systems, and Process Equipment changes. The New Chemical Acquisition Process (NCAP) is utilized for introducing new chemicals to a site, including obtaining all necessary approvals based on the hazard class and risk associated with a chemical. The Laboratory Hazard Analysis (LHA) is a process for evaluating hazards of laboratory experiments (non-PSM and non-equipment changes) through a team based risk assessment process. The Digital Control Systems Hazard Analysis (DCSHA) is a tool for evaluating changes specific to laboratory control and display systems. The Equipment Change and Assessment Process (ECAP) focuses on evaluating hazards associated with all equipment changes. Finally, a New Employee Orientation Passport process was designed to assist with onboarding personnel new to a site or work group. In support of the new strategic alignment, a new software tool was developed in a collaboration between researchers, corporate environmental health, safety, and security, and information technology employees. The new streamlined process eliminated paper forms and improved workflow efficiency through a variety of new tools. Some distinct advantages included the move to an all e-signature platform and the involvement of resources throughout the organization only when necessary. The program operates on a risk assessment basis and focuses on increasing the level of questioning and evaluation as the associated risk of activity or personnel development increases. In all the modules, a series of core/guiding questions are answered and the dynamic character of the program responds by modifying further questions or the level of review required. The program has now been implemented for three years and has proven very successful in addressing the needs of the research and development organization. Several awards have been received as a result including a Best Practice Award during OSHAs Voluntary Protection Program (VPP) audit.

512058 Developing a Pilot Plant Safety Culture within an Historically Laboratory Focused Work Force

Abstract:

While both pilot plants and laboratories are smaller than most chemical manufacturing facilities, they still present occupational and process hazards. They both also offer different challenges compared to one another. Hazards of dealing with new technologies are similar, but differences arise due to their size, the environments they are housed in, and compliance standards.

As our company grew, it made a step change by expanding from laboratory research and engineering services and added extensive pilot plant operations in a short period of time. These operations introduced a new level of hazards and required a different type of thinking, but the organization was still using laboratory methodologies to evaluate hazards and still had a laboratory mindset regarding operational procedures and protocols.

The pilot plant was staffed with people of varied backgrounds. Some had been at the company for several years and had worked in the research laboratories, and some were new hires specifically brought in to staff-up the new pilot plant area. Some of the technicians also had extensive prior pilot plant experience in the very same building but within a different type of company with a different focus and level of support from outside organizations. Some came with commercial operations experience, and some were completely new to the industry. At the same time, the company was also experiencing an influx of leadership with backgrounds in commercial scale chemical operations. This mixture of new operations, new people, existing people taking on new roles, new leadership styles, and simply new expectations, had the potential for negative consequences. While there have been hurdles to overcome and the work processes continue to evolve, the organization has thus far risen to the challenge.

This presentation will explore the challenges that the organization faced and the cultural shift and work processes that were required to meet the new expectations and demands. This includes topics such as developing occupational safety work processes, adjusting the management of change process, encouraging new approaches to process safety, and overcoming undesired behaviors and perceptions. The presentation will also discuss how the organization is continuing to evolve five years later as the pilot plant businesses continue to grow and we make continuous improvements to our occupational and process safety systems.

505848 Relief Design for Laboratory and Pilot Plants

Abstract:

Relief design at the laboratory and pilot scale presents unique challenges when compared to full commercial plants. Properties used for relief device sizing, commercial availability of devices and methods of effluent disposal can make things difficult. Furthermore, design codes typically applied to commercial plants may or may not apply. This session will touch on some of these issues, and how they are typically handled.

512849 Computational Tools for Designing Innovative Chemical Products

Abstract:

Designing new, innovative chemical products has always been a challenging task. Traditionally, the design process would require decades of product specific knowledge and a considerable amount of trial and error experimentation. Today we are developing a more systematic, scientific approach to chemical product design. New software tools can generate tens of thousands of candidate molecular structures and mixture formulations. Computational structure property relationships can be used to quickly predict the properties of each candidate and screen them against constraints on performance, safety and environmental impact. Promising candidates can then be screened using additional tools to ensure they can be produced in an efficient and sustainable manner.

We will discuss this new approach to chemical product design, demonstrate the use of computational tools for screening candidates and detail several examples. Specifically, we will:

1) Present an overview of the concepts behind computational chemical product design algorithms detailing how the computer can generate and enumerate â €œcorrectâ€ molecular structures.

2) Explain how group contribution estimation techniques are used to screen candidate chemicals.

3) Show how mixture estimation techniques can calculate the physical properties of candidate formulations.

4) Demonstration how basic chemical engineering knowledge is used to transform the often abstract goals of chemical product design into sets of well defined chemical and physical property constraints.

We will present several product design examples including the design of new refrigerants, solvent entrainers for extractive distillation, blends of lubricating oils, solvents for ink-jet printer ink and propellants for shaving cream.

513107 The Effect of Feed Compositon Variations from Upstream Plants on the Downstream Design Configruation of Natural Gas Liquid (NGL) Recovery Processes

Abstract:

Gas processing plant is subjected to variations in upstream conditions that has an impact on the performance of the in-operating plant. The feed composition determines what types of process configuration has to be implemented for the process. In this study, six representative NGL recovery processes were selected based on the criteria of their commercial availability, operating and capital cost savings. Each process scheme were built using HYSYS under common operating conditions for the purpose of comparison. The developed process schemes were tested for a range of various feed compositions classified as lean and rich. Economic evaluation were made by measuring the total annualized cost (TAC). The enhanced NGL recovery process (IPSI-1) showed the lowest TAC value compared to the others due to the presence of the self-refrigeration that reduces the external usage of propane refrigerant for the process. The trade-off that exist between the profitability and reliability of holding the process operating constraints were also been evaluated. The IPSI-1 scheme showed the highest profit margin with less than 6% risk of violation for the constraints.

513512 Modeling and Simulation of Diesel Hydrotreaters for ULSD Production: Selection of an Optimal Catalyst Composition

Abstract:

The required specifications for ultra-low sulfur diesel (ULSD) have driven refiners towards high performance hydrotreating catalyst that can efficiently remove difficult sulfur down to 15 ppm or lower in the final product. To select the best catalyst composition for the design of grassroots facilities or the revamp of existing ones, engineers need to account for the type of feed and the optimal operating conditions to reach hydrodesulfurization (HDS) and cetane number improvement targets.

This work focuses on the development of a flexible model for commercial ULSD hydrotreating units that predicts product properties, yields and hydrogen consumption for different catalyst compositions. The model uses a species set that contains the necessary structural information required to characterize different types of sulfur species as well as to distinguish among mono-/poly-aromatics, naphthenes and naphtheno-aromatics to model relevant reactions. The proposed reaction set accounts for desulfurization pathways by hydrogenolysis and hydrogenation, saturation of poly-aromatic, mono-aromatic and olefin species, cracking, and inhibition effects by nitrogen impurities in the feed.

Experimental data from the literature for CoMo and NiMo catalysts at different levels of desulfurization and aromatic saturation with temperature, pressure, and LHSV were used to determine two sets of kinetic parameters that represent each type of catalyst. Case studies were developed to assess the performance of each individual catalyst and combinations of catalysts in commercial ULSD hydrotreaters under different scenarios covering HDS and cetane improvement with limited hydrogen consumption, increasing amounts of light cycle oil in a straight-run gas oil feedstock blend, and low to high pressure operation from start of run to end of run conditions. For each scenario, an optimal catalyst composition was found, constrained to a final product sulfur specification of 15 ppm.

513598 Optimal Reaction Design of Alkoxy Silane Copolymers Using Design of Experiment Methods

Abstract:

Silicone-based copolymers such as polysiloxane and polysilsesquioxane have unique properties that make them well-suited for many industrial applications such as encapsulation and packaging of electronic components, anti-microbial coatings, and microelectronics[1,2]. They are particularly used in multi-layer lithography, a type of fabrication process for integrated circuits that provides increased resolution and sensitivity compared to other methods, and which relies on silicon-containing polymers to enable sharp pattern transfer. Integrated circuit applications require materials with very precise and tight quality specifications. To obtain these specifications, a good quantitative understanding of the molecular structures and properties of organosilicon polymers[3], which undergo structural changes during processing[3], is needed. Thus, understanding how defects in molecular domains influence the properties of the silicon-containing copolymers enables high-quality and controlled production scheme.

In this study, we applied the design of experiment methods (DOE) to gain a systematic understanding of the effect of different factors such as acid catalyst concentration, water to alkoxy silanes ratio, alkoxy silane monomers ratio, and condenser flow rate on the copolymer physical properties and structural chemistry. Here, poly(MTMS-co-TEOS), with the range of weight average molecular weight of 3000-6000 $gmole^{-1}$, was synthesized. The effect of reaction parameters and their interactions were studied through using different characterization methods such as TGA-IR, AFM-IR, NMR spectroscopies to achieve the optimal reaction condition for manufacturing products within the specification limits. It was found that increasing the ratio of condensation rate to hydrolysis rate and using a high ratio of trimethoxymethylsilane (MTMS) to tetraethoxysilane (TEOS) monomers induce a transition in the molecular structure and favors formation of random and cage structures, which can be attributed to defects, to linear ones. Moreover, we found out that acid addition rate has the strong influence on the formation of defects. We concluded that the formation of random and cage structures can be suppressed through increasing the condensation rate, improving the mixing rate, and controlling the addition rate of the acid catalyst.

References:

(1) Okoroanyanwu, Uzodinma. "Chemistry and lithography." Bellingham, WA: SPIE, 2010.

(2) Schmid, H., and B. Michel. "Siloxane polymers for high-resolution, high-accuracy soft lithography." *Macromolecules* 33.8 (2000): 3042-3049.

(3) Reichmanis, Elsa, et al. "Organosilicon polymers for microlithographic applications." *Advances in chemistry* 224 (1990): 265-281.

513189 Holistic Analysis of Emissions from a Small, Modular Fast Pyrolysis System for Conversion of Biomass and Mixed Waste

Abstract:

Total emissions for a small-scale (45 kg/hr, nominally 1 ton/day), transportable pyrolysis system, including on-site utilization of the fuels and byproducts, were determined from measured burner emissions using pyrolysis bio-oil, char, and gas fuels derived from mixed solid waste (MSW) and biomass. Fast pyrolysis provides an alternative to open burn pits that have been used at remote bases, and emit harmful gaseous emissions and particulate matter. By comparison, fast pyrolysis can convert MSW into useful heat and power while producing sterile ash and relatively clean combustion products. Combustion of pyrolysis byproducts (i.e., char and gas) was required to generate process heat for drying the feedstock to the required moisture content (<10%) and to achieve the desired volumetric waste reduction target (>95%). Utilization of byproducts was especially important for these small-scale systems where heat losses and parasitic power draws became more challenging as the surface-to-volume ratio of the unit operations increased. Emissions were measured from a process gas burner, bio-oil burner, and char burner that compose the fuel utilization section of a pilot-scale pyrolytic solid waste remediation system (*pyâ*€'SWRS), and the emissions were compared to the relevant standard set by the U.S. Environmental Protection Agency (EPA). The gas burner was evaluated using propane (LHV of 47 MJ/kg) and a surrogate process gas (LHV of 8 MJ/kg) with a molar composition of 16% CH_4, 39% CO_2, and 45% CO. Several 80% (m/m) ethanol and 20% (m/m) bio-oil fuel blends were prepared to measure the CO_2, CO, NO_x, THC, and PM emissions of a heated, air-atomizing bio-oil burner. Fast pyrolysis char produced from pine feedstock was combusted with a pilot-scale fluidized bed reactor to demonstrate the feasibility of process heat generation. In the char burner, the fluidizing gas flow rate, gas inlet temperature, excess air, and reactor temperature were varied to determine the effects on emissions and combustion efficiency. Based on the emissions of each burner, the total emissions for CO, NO_x, SO_2, and total PM were determined for the pilot-scale *pyâ*€'SWRS and compared to the EPA standard for other solid waste incinerators (OSWI), which regulates nine pollutants. Of the four common pollutants, CO, NO_x, SO_2 met the EPA OSWI standard, while total PM (32.6 mg/dscm) was slightly over the 30 mg/dscm limit. The bio-oil burner contributed the most to the total PM emissions, based on the tested burner conditions. We postulate that the PM could be reduced below the limit by tailoring the fuel-air equivalence ratio of the bio-oil burner. The other five pollutants included in the OSWI standard (dioxins, cadmium, lead, mercury, and HCl), which depend primarily on the feedstock composition, were also considered. The total emissions data were fed into a decision support model for waste remediation at remote locations, which considered environmental compliance as one of the performance metrics.

512939 Modular System for Converting Lignocellulose into Pyrolytic Sugars and Phenolic Oil

Abstract:

Easy Energy Systems and Iowa State University are collaborating on the design, construction, and demonstration of a 50 ton per day pyrolysis system based on modular manufacturing principles. The goal is to commercialize a system suitable for processing distributed resources of lignocellulosic biomass into pyrolytic sugars and phenolic oil. Success is premised on intensification of several processes and designing major unit operations and subsystems as modules that fit into standard ISO shipping containers. The modules are field assembled into a complete biomass processing plant at scales much smaller than once imagined for commercial production of cellulosic biofuels and biobased chemicals.

Processes to be intensified include: provision of energy to the pyrolysis process through autothermal operation; production of sugars through thermal deconstruction of plant polysaccharides; and anaerobic digestion of the aqueous phase produced during pyrolysis. In contrast to conventional pyrolysis in which oxygen is excluded from the reactor, autothermal pyrolysis adds air at equivalence ratio less than 10% (considerably lower than autothermal gasification, for example) to provide the enthalpy of pyrolysis without the need for external heat transfer equipment, greatly simplifying system design. Autothermal operation removes the heat transfer bottleneck in fast pyrolysis, increasing biomass throughput by several fold compared to conventional pyrolysis in a similarly sized reactor. Pyrolysis sugars are levoglucosan (a dehydrated form of glucose) from cellulose and principally xylose from hemicellulose. Although sugar yields from pyrolysis of lignocellulosic biomass is normally very low, a simple pretreatment prior to pyrolysis enhances sugar production several fold. The sugar is recovered as a syrup contaminated with phenolic monomers. The syrup is cleaned with a resin adsorption filter, which allows its use as fermentation substrate. The phenolic oil, derived from lignin in the biomass feedstock, has several applications. It can be cured to a solid known as Lignocol, which has applications as a renewable fuel substitute for coal in electric power generation. It can also be emulsified with diesel fuel or dissolved in ethanol to produce a low-viscosity fuel oil suitable for small-scale power generation. With suitable hydroprocessing the phenolic oil can also be converted into renewable gasoline and diesel fuel. The aqueous phase is anaerobically digested to bio-methane after removal of fermentation inhibitors.

This presentation will describe pilot-scale tests and design activities to support construction of the 50 ton per day demonstration plant in 2018.

513197 Anaerobic Treatment of High Strength Industrial Wastes By Moving Bed Bioreactors (MBBR)

Abstract:

Anaerobic Treatment of High Strength Industrial Wastes by Moving Bed Bioreactors (MBBR)

Somnath Basu
Headworks International
11000 Brittmoore Park Drive, Houston, TX 77014
sbasu@headworksintl.com

ABSTRACT
Effluents from many industrial processes contain high concentrations of organic wastes (COD). These include petrochemical, food & beverage, and breweries & distilleries among others. Treatment of high COD wastes pose major challenges to the operating cost and environment traditional aerobic activated sludge process due to their:
a) high energy cost of aeration as every kilogram of COD requires one kilogram of oxygen; and
b) high cost of transportation and disposal of residual solids resulting from biochemical reactions.
These two activities constitute the largest operating cost of a wastewater treatment plant.
Anaerobic processes are net energy positive and sustainable for high strength waste treatment compared to their aerobic counterpart as:
a) they do not consume any energy for aeration as these processes take place in absence of air;
b) very low generation of residual biosolids requiring much lower hauling and disposal costs.
On the other hand, anaerobic processes produce biogas that can be utilized as a source of energy for heating, steam and power generation on-site. From energy viewpoint, the production far outweighs the consumption to maintain the bioreactor well mixed by agitators and at 35 to 38 C. Thus, anaerobic processes are sustainable as biogas is a source of 'Green Energy'.
Despite these advantages, anaerobic treatment faces a significant operating challenge in solid-liquid-gas separation can be challenging because:
a) trapped gas impedes solids settling
b) waste sludge does not flocculate well
This difficulty leads to occasional washout of biomass from the reactor. Restart and stabilization of reactor can be time consuming and can affect the smooth operation of production processes.
An anaerobic MBBR reactor overcomes this constraint as the biomass in the reactor is attached to media surfaces while the media themselves are freely suspended and uniformly distributed within the reactor. Other advantages of this type of reactor configurations, as listed below.
• Robustness against toxic or shock loads;
• Resiliency to peak flows and loads; and
• Easy recovery from washout situations as attached growth provides an inventory of biomass
This paper will present the fundamentals of anaerobic MBBR as a sustainable process for high strength waste treatment with the results of a case study of a pilot scale brewery wastewater treatment.

512935 Best Practices and Case Studies of Managing Process and Operational Safety in Process Development from Laboratory to Pilot

Abstract:

In addition to the potential impacts on health of staff and colleagues, preservation of R&D facilities and license to operate, nothing derails a new process development more quickly than a safety incident or even a â€œnear missâ€ with substantial potential consequences. Managing safety risk in process development is a key cornerstone for success, and sets the standard for how process safety at the commercial scale will be managed. Challenges include the consideration of multiple routes and process chemistries, including multiple new materials (catalysts, solvents, adsorbents) whose properties may not be fully known. Extra care must be taken in design of bench and pilot facilities to accommodate unknowns, and changes in process line-ups and chemistries that will inevitably be required to identify viable and optimal commercial designs. In some cases, inherent safety risks may drive the selection of a given pathway over an alternate route with increased risk.

This presentation will cover a range of case histories in process development safety management, including the advantages of small scale testing during early screening, use of barricades to facilitate testing and redesign of oxidation processes, providing for added protection against personnel exposures, challenging existing protocols to improve safety before expanding a program with highly reactive chemistries, making correct decision in safety management despite fast-track commercial pressure. The importance of the process development engineer in leading process safety by design and by example behavior, is emphasized.

511931 Implementing Laboratory Safety at Third-Party Laboratories

Abstract:

For an independent testing laboratory, customers can range from major petrochemical companies to independent lubricant manufacturers to private individuals interested in one-off testing. This diverse customer base results in a range of materials to be tested as well as inconsistent documentation and sample handling; both factors contribute to the unique safety challenges third-party laboratories face. Two tests – the Hot Liquid Process Simulator (HLPS) and ASTM D2879, Vapor Pressure by Isoteniscope – will be used as "case studies" to demonstrate how independent laboratories can mitigate safety risks. Discussion will include how staff training, sample intake and handling procedures, laboratory infrastructure, and even test instrument design can be employed in a manner that minimizes health and safety risks.

Systematic Approaches to Mitigate Hazards in the Design and Operation of Experimental Systems

Andrew Kasick [Presenter], Sunggyu Lee, Sustainable Energy and Advanced Materials (SEAM) Laboratory, Ohio University, Athens, OH

Introduction

The State of Safety in the Academic Research Laboratory

A clichéd observation, which arises frequently in the world of the chemical sciences, is that safety practices in academic research settings are woefully behind the practices in industrial, commercial, or governmental settings. Beyond personal anecdotes, this divergence in attitudes and practices has received critical professional and administrative evaluation over the last decade. In 2012, Dow Chemical began a collaboration with a number of universities, including the University of Minnesota, Pennsylvania State University, and University of California, Santa Barbara, to gauge how Dow, as a research supporter, could take a more active role in promoting safety at the universities with whom they were partnering (*1*). In a 2012 summary report on opportunities for improvement in graduate and postgraduate education in the chemical sciences, the American Chemical Society (ACS) emphasized the practical and professional importance of academia's adoption of stronger safety practices and culture, more in line with those encountered in industry (*2*). This should demonstrate that adopting professional safety practices in academia goes beyond averting incidents in research laboratories, because the inherent ability to carry out research in a safe manner is a skillset valued in professional chemical science settings, and graduates should begin the development of this skillset in an academic setting.

The recognition that academic laboratory safety attitudes and activities fall critically short of those in industrial, commercial, and governmental facilities should serve as an invitation to elevate the former's standards and expectations regarding safety in research activities. Organizations across the breadth of the chemical sciences can play important and distinct roles in these efforts. Earlier, it was mentioned that Dow was interested in taking an active role in promoting safety in the universities with whom they were collaborating; to this end, in 2013, Dow launched a website featuring dozens of safety videos, in an effort to provide resources to other institutions looking to improve safety (*3*). Another example of an outside agency promoting safety within a collaborating organization is found in the U.S. Department of Energy's (U.S. DOE) Fuel Cell Technologies Program, which in April of 2010 issued a document to provide current and potential collaborators with information on planning and performing research on hydrogen production, storage, and distribution in a safe manner (*4*). In a Case Study issued by the U.S. Chemical Safety and Hazard Investigation Board (CSB) in 2011, in response to an accident at a University the previous year, the potential benefit was noted of granting agencies taking an active role in observing and promoting safety laboratory practices within collaborating academic institutions; the above examples from Dow and the U.S. DOE demonstrate this principle in action (*5*). The observation by the CSB and the examples of Dow and the U.S. DOE should demonstrate the positive development that can be made when academic research groups draw from models outside academia, in this case industrial and governmental, in developing the internally followed safety practices.

The Mini-Pilot Plant in the Laboratory

Many critiques of academic laboratory practices have focused on matters of proper chemical hygiene and personal protective equipment (PPE) use, which are extremely important and relevant to all laboratories handling chemicals. It is important, however, to recognize that academic laboratories, due to unique and diverse areas and methods of research, will encounter forms of hazards that run far afield of traditional chemical handling concerns. One such form is the use of experimental chemical process systems, or bench scale mini-pilot systems, in academic research, which go beyond the typical wet chemistry hazards and introduce risks associated with high temperatures and pressures, chemicals handled in physical states that elevate their dangers, material compatibility and corrosion issues, in addition to electrical and mechanical hazards.

Generally speaking, the hazards introduced by experimental reactor systems within academic research laboratories are closer to those found in industry than in traditional academic laboratories. Drawing from this observation, work has been done to apply industrial safety practices to the construction of a batch hydrogenation experimental system under the leadership of Dr. Sunggyu Lee at the Sustainable Energy and Advanced Materials (SEAM) Laboratory at Ohio University, and to develop a generalized procedure to incorporate these practices into future work in the Laboratory. Only a few accounts have been published demonstrating the utility of applying such established techniques in pilot plant and research laboratory settings (*6,7*). But the experience at the SEAM Lab is unique in that the easily identifiable manner in which these practices directed design and construction decisions to obtain safety benefits, and this presentation advocates for the adoption of such practices across the academic research community by emphasizing the accessibility of the required tools and techniques.

Industrial use of pilot plants differs from systems run in a laboratory setting regarding size and purpose, but the use of the term "pilot plant" or "mini-pilot plant" is warranted in the bench-scale, academic context. Noting that some technical definitions do not restrict pilot plant identity to size limits, such as the National Fire Protection Association's NFPA 45 definition, on a more fundamental level, a laboratory system that has been constructed for repeated and/or continuous experimental runs of a chemical process, with ready alteration of process variables by the operator, should be viewed from a safety perspective as a bench-scale pilot plant, or mini-pilot plant (*8*).

This is a particularly important perspective to adopt when considering the SEAM Lab at Ohio University. In this 16,000 ft^2 facility, a wide variety of alternative fuels, advanced polymers, and specialty technologies research is carried out with custom designed and built experimental process systems. Specific research interests in the SEAM Lab required the design and construction of a batch hydrogenation experimental system; hydrogenation reactions have an inherent variety of hazards associated with them, including the flammability issues of hydrogen gas, the elevated pressures and temperatures at which the reactions occur, and the typically reactive and pyrophoric nature of the catalysts used (*9*). The recognition of these hazards, along with the growing commitment within Ohio University's Russ College of Engineering and Technology to elevating the safety of research activities, motivated a thorough assessment of how the hydrogenation unit could be constructed and operated to mitigate the involved hazards, and how this experience could serve to elevate SEAM Lab practices as a whole. As such, this presentation will detail the ongoing efforts to analyze the hazards and risks inherent within the research and introduced by the system's design, to incorporate this more sophisticated understanding of the hazards and risks into the design and selection of suitable elements of the system, and to communicate the skills developed in this particular endeavor to the other members of the research group, so that safety practices as a whole are elevated. Before delving into the discussion of the specific work relating to the hydrogenation system's design and construction, it is important to overview the Russ College's efforts to promote an attentive attitude towards safety in research activities, which contributed to the context from which this work was motivated.

The Russ College Safety Plan

Around the same time that serious consideration was going into how the hazards introduced by the hydrogenation unit could be mitigated, the researchers at the SEAM Lab were introduced to efforts within Ohio University's Russ College to formally document the research activities carried out within the College. Going by the name the "Russ College Safety Plan" (RCSP), a framework would be provided by which researchers can catalog the design of research operations and experimental scope, define the standard operating procedures and conditions (SOPs and SOCs) that are within the bounds of the activity's intended design, and evaluate the potential impact of an accident via a Failure Mode and Effects Analysis (FMEA) (*10*). While at present this is not a mandatory requirement within the College, its application of a FMEA to research activities inspired a new attitude, concerning how tools and resources used in industry could be applied within the Laboratory. Beyond documenting a research activity, an enhanced level of safety can be achieved by actively applying hazard evaluation tools to inform design and operation decisions. The first step in applying this outlook was in assessing and evaluating the hazards and risks associated with performing hydrogenation reactions in the experimental unit.

Body

The following description represents the major topics covered in this presentation. The presentation itself will feature appropriate supplemental information, such as photographs, hazard evaluation tool examples, and descriptions of the construction process, which will reinforce the concepts described in this extended abstract.

Hazard Evaluation

Potential safety concerns cannot be adequately addressed without first being identified, and a number of tools and techniques exist to encourage a thorough analysis of safety concerns within a system or activity. During the earliest efforts in building the hydrogenation system, the identification of hazards was not approached in a deliberate manner, and as a result potential safety concerns arose somewhat unexpectedly, and the progression of the work was sidetracked by efforts to research and address the encountered concern. Admittedly, this was not an efficient practice in either identifying hazards or maintaining progress on unit construction, but with this recognition efforts were made to research how hazard evaluation efforts are carried out in the construction of industrial pilot plants and commercial chemical facilities, for the purpose of informing construction activities within the Laboratory.

The first such application of this knowledge came with an area-based "What-If" analysis, which adopted a "Gedankenexperiment" approach in which hypothetical situations were proposed, and potential ramifications were considered. The What-If analysis tool that has previously been identified as particularly well suited for the construction of pilot plants and new chemical processing technologies (*11,12,13*). Areas in which the what-if analysis identified previously unacknowledged hazards included potential overpressure situations arising from regulator failure and the impacts that an electrical utility failure could have on unit operation. This represented an effective effort in uncovering hazards that may have been overlooked, or may have been realized at a later point when responding in a proper manner would have been more difficult; the concern that other hazards still were being overlooked motivated the application of a more structured review of the proposed process. To this end, a hazard and operability (HAZOP) analysis provided direct attention to a much wider range of abnormal process situations (*11,12.13,14*). Finally, a consideration of the likelihood and potential impact of the hazards previously identified was a critical step in evaluating how work should be prioritized to respond to the most pressing hazard concerns. To this end, the hazards identified in the HAZOP analysis was augmented with probability and severity rankings, and these rankings can assist in placing the hazards along a risk matrix, a valuable tool in visualizing how the hazards should be prioritized for mitigating action (*12,14*).

These studies represent important documentation and record keeping about the system being built, which are important practices for enhancing system safety and loss prevention. As work continues and potential hazard sources are identified, they can be incorporated into this documentation to promote ever-improving safety practices. With a more thorough understanding of the potential hazards of the hydrogenation system, the strategies for addressing hazards in the design and selection of suitable elements of the system was needed. Once again, this undertaking was promoted by an understanding and application of the codes and classifications that are observed in industry, put poorly understood in academia.

Hazard Mitigation and Management

The combined knowledge of the design and design philosophy of the system and the specifications of the hazards that needed to be managed provided a basis for specifying the equipment suitable for installation in the process. Elements such as material compatibility, pressure handling and relief capability, and electrical area classification were some of the characteristics which need to be factored into component selection, based on the relevancy of the earlier hazard evaluation to the given component (*13*). For a researcher communicating with vendors to purchase a piece of equipment, a lack of familiarity with how such considerations guide the selection process can make the task confusing, and potentially lead to errors. Once again, studying industrial and commercial practices provided vital information in navigating this critically important task. For example, the National Fire Protection Association's NFPA 70, the National Electric Code (NEC), details widely recognized safe practices in selecting electrical components for a given application. In the case of the experimental hydrogenation system, the standards for area classification in NEC Article 500 were applied wherever the opportunity presented itself to the system in the selection of the electrical components and housings, such as the safety shutoff valve for the hydrogen feed line and the combustible gas detector housing (*15*). The concerns identified in the hazard evaluation of a proposed experimental pilot plant system were addressed in the design and construction stages by interpreting and applying the codes and regulations followed in industry. This reflects a large opportunity for improvement within academic institutions, because the utilization of codes and regulations can be impacted by a lack of awareness and access to the relevant material. Despite the intricacies of resources such as the NEC guidelines, a newcomer with a basic understanding can still improve the decision-making process in the academic research setting based on such material. The complexity of hazard evaluation and equipment specification, however, can be difficult in the academic setting, where personnel may have limited prior experience in applying these tools to practical situations; for this reason, the work on the hydrogenation system has served as the basis for formal and informal efforts within the SEAM Lab to promote broader understanding and consciousness about these matters, so this improved skillset can be applied to promote safety in a wider range of laboratory activities and future professional activities.

Promoting Lab-Wide Safety Practices

The expectation of the above described efforts, to identify and effectively address hazards in the construction of the SEAM Lab's hydrogenation unit, had the very definite goal of allowing researchers to collect data and achieve research objectives without encountering a hazardous incident, a circumstance that could delay or end research activities, damage equipment, or cause physical injury or death to laboratory personnel. Given the range of research performed in the SEAM Lab using prototype chemical processing units, the lab-scale pilot plants, the efforts devoted to the hydrogenation unit should also have a goal of elevating the safety quality and consciousness among the researchers within the Laboratory, so that future safety concerns can be evaluated and addressed in a more robust and efficient fashion.

As work on the hydrogenation system has progressed, attention was turned to implementing formal and informal means of advancing toward the goal of generalizing the lessons learned and work devoted to building the unit. Formally, disseminating this knowledge involved making flowcharts and questionnaires that assist in interpreting relevant code descriptions. Informally, the incorporation of unique safety considerations and the analysis of safety incidents at other academic institutions became a regular feature of weekly group meetings, so that a realistic attitude about potential laboratory hazards developed within the research group. The expectation was that the informal promotion of safety consciousness within the Laboratory would enable the group members to be proactive in assessing the need for hazard evaluation and mitigation in research activities, and the formal instructions and guidelines would allow personnel to quickly locate the information necessary to answer questions and make informed decisions relating to safe research activities. This will benefit the Lab by promoting knowledge and access of common codes, regulations, and resources, and benefit the hydrogenation project by establishing a detailed history of the development of the system and its operating protocols.

Conclusion

There is a growing attentiveness across the chemical sciences that research activities in academic institutions would greatly benefit from the adoption of professional safety practices encountered in industrial, commercial, governmental, and other settings. The promotion of safety practices would not only reduce the potential for damaging incidents in academic laboratories, but would produce graduates better suited for professional settings and careers. For some laboratories, such as the Sustainable Energy and Advanced Materials (SEAM) Laboratory at Ohio University, research is carried out in mini-pilot plant experimental reactor systems, which approach full-scale pilot plants in the safety considerations they demand. To truly elevate academic safety practices relating to experimental process systems and pilot plants, tools and practices can be drawn from industry. This presentation has demonstrated such an undertaking in the design and construction efforts devoted to a batch hydrogenation experimental system, as shown in the efforts to identify and evaluate the hazards introduced by the system, in the measures taken to mitigate the evaluated hazards in the system's design and element selection, and in the efforts taken to generalize this experience, so that SEAM Lab practices as a whole can proceed with a high degree of professional safety. This reflects a graduate student's efforts to improve the integrity of a specific research project in a specific Laboratory, drawing motivation from prior work experience and the example set with the development of the University's Russ College Safety Plan. It is, however, part of a bigger professional shift in the knowledge, attitudes, and practices regarding academic research. In the last decade, attention across the field of chemical sciences has been drawn to promoting of safety consciousness and better practices in academic laboratories, from professional institutes and associations to research funding agencies, both industrial and governmental. Whatever position within the field of chemical sciences an individual or organization occupies, the potential exists to promote this shift.

Acknowledgments

First and foremost, thanks must be extended to Dr. Sunggyu Lee, the Principle Investigator of the Sustainable Energy and Advanced Materials Laboratory. His vision and hard work originated the hydrogenation project and enabled its drive forward, and his support for the described hazard evaluation and mitigation efforts demonstrate the critical role a PI plays in overseeing research activities. Many thanks are owed to Dr. Damilola Daramola, for his tireless efforts in managing the daily operations of the Lab, and the wide breadth of research activities which occur there. Credit is owed to the Russ College Safety Coordinator Shyler Switzer and Dr. Greg Kremer, who have worked for a decade to develop and refine the Russ College Safety Plan (RCSP). The RCSP influenced the proactive attitude adopted in building the hydrogenation system, and should be recognized as an important step in the above mentioned professional shift regarding academic research safety. Finally, thanks are in order to the fellow members of the SEAM Lab who witnessed and assisted in these efforts over the last seven months: graduate students Bahareh Baheri, Matthew Conley, Amy Lindenberger, and Lujie Ye and undergraduate students Michael Cole, Morgan Schroeder, Haley Vega, and Dillon Wiesler.

References

1. Kemsley, J. Dow Chemical Teams Up With Universities On Laboratory Safety. *Chem. Eng. News* [Online] **2012**, *90* (44), 21-23. https://cen.acs.org/articles/90/i44/Dow-Chemical-Teams-Universities-Laboratory.html (accessed January 23, 2018).
2. Banholzer, W.F., et al. *Advancing Graduate Education in the Chemical Sciences: Summary Report of an ACS Presidential Commission* [Online]; ACS Presidential Commission; American Chemical Society: Washington D.C., 2012. https://www.acs.org/content/dam/acsorg/about/governance/acs-commission-on-graduate-education-summary-report.pdf (accessed February 20, 2018).
3. Kemsley, J. Dow Unveils Lab Safety Website. *Chem. Eng. News* [Online] **2013**. https://cen.acs.org/articles/91/web/2013/05/Dow-Unveils-Lab-Safety-Website.html (accessed January 23, 2018).
4. *Safety Planning Guidance for Hydrogen and Fuel Cell Projects* [Online]; U.S. Department of Energy Fuel Cell Technologies Program: Washington D.C., 2010. https://www.energy.gov/eere/wipo/downloads/safety-planning-guidance-hydrogen-and-fuel-cell-projects (accessed March 25, 2018).

5. *Case Study: Texas Tech University Laboratory Explosion* [Online]; No. 2010-05-I-TX; U.S. Chemical Safety and Hazard Investigation Board; Washington D.C., 2011. https://www.depts.ttu.edu/vpr/integrity/csb-response/downloads/report.pdf (accessed February 10, 2018).
6. Zakeski, J. Improving engineering research laboratory safety by addressing the human aspects of research management. *J. Chem. Health Saf.* [Online], **2009**, *16* (3), 5-20. OhioLINK Electronic Journal Center. DOI: 10.1016/j.jchas.2008.06.001 (Accessed February 11, 2018).
7. Aziz, H.A.; Shariff, A.M.; Roslan, M.R. *Managing Process Hazards in Lab-Scale Pilot Plant for Safe Operation. Am. J. Eng. App. Sci.* [Online], **2012**, *5*, 84-88. http://thescipub.com/pdf/10.3844/ajeassp.2012.84.88 (Accessed January 29, 2018).
8. NFPA 45: Standard on Fire Protection for Laboratories Using Chemicals, 2015 Edition. *National Fire Protection Association National Fire Code Series*; National Fire Protection Association: Quincy, MA, 2014; 45-10.
9. National Research Council of the National Academies, Committee on Prudent Practices in the Laboratory: An Update Board on Chemical Sciences and Technology Division on Earth and Life Studies. *Prudent Practices in the Laboratory: Handling and Management of Chemical Hazards*; National Academies Press: Washington, D.C., 2011; pp 70, 74, 129-130, 135-136, 171, 224.
10. Switzer, S. Ohio University, Athens, OH. Personal communication regarding the Russ College Safety Plan, 2017.
11. Palluzi, R.P. *Pilot Plant and Laboratory Safety*; McGraw-Hill: New York, 1994; pp 43-45.
12. American Institute of Chemical Engineers, Center for Chemical Process Safety. *Guidelines for Hazard Evaluation Procedures*, 2nd ed. With Worked Examples; Center for Chemical Process Safety of the American Institute of Chemical Engineers: New York, 1992; pp 48, 117-122, 131 -159, 198-210.
13. McConville, F.X. *The Pilot Plant Real Book: A Unique Handbook for the Chemical Process Industry*, 2nd Ed.; FXM Engineering & Design: Worchester, MA, 2007; pp 1-2, 1-12 – 1-13, 2-4, 2-14, 5-2, 5-7 – 5-9, 8-21.
14. Mitchell, K.J.; Longendelpher, T.M.; Kuhn, M.C. *Kenexis: Safety Instrumented System Engineering Handbook*; CreateSpace Independent: Middletown, DE, 2010; pp 9-18, 25-31, 37-40.
15. NFPA 70: National Electrical Code, 2017 Edition. *National Fire Protection Association National Fire Code Series*; National Fire Protection Association: Quincy, MA, 2016; 70-349 – 70-355.

511441 Where Does Laboratory Safety Start ?

Abstract:

Industrial laboratory safety starts in university laboratories. If industryâ€™s raw material; i.e., new hires, have been introduced to an OSHA-style safety program at university, then industry will find safety indoctrination of new hires easier to accomplish.

However, safety culture in university laboratories encounters many significant issues. First, many universities are state bureaucracies, which do not receive the same level of scrutiny from another State or Federal bureaucracy as does industry. Second, today, many university students come from countries where safety is not a high priority: having a job or producing products are higher priorities than safety in many regions of the world. Third, graduate students are far more interested in obtaining experimental results than learning and enforcing safety regulations. Fourth, the university laboratory staff confronts a new group of students every semester, which requires safety orientation and indoctrination from â€œground zeroâ€.

This presentation discusses how the authors have resolved these issues with regard to undergraduate laboratories at a state university and a private university.

513165 Process Safety and Scale-up Strategy for Coating Polymers

Abstract:

Process safety is key part of scale-up from lab to pilot to production. A comprehensive scale-up strategy that encompasses each step will be presented. Calorimetry and its importance at validating reaction kinetics, vent sizing, and exothermic reaction modeling will be discussed. Value of this key information in understanding the thermal safety of chemical processes, risk assessment, and process design as part of the scale-up protocol will be outlined. Design, development, and scale-up of phenolic resins will be given as a case study. The approach will outline the significance of modeling, design of experiments, latitude studies, pilot batches, to make data driven decisions as part of the commercialization process.

514236 Integration of Smart Manufacturing in the Advancement of Polymer Processing

Abstract:

In lieu of improving efficiency and embracing smart manufacturing technologies, the U.S. Department of Energy (DOE) has instituted the Clean Energy Smart Manufacturing Innovation Institute (CESMII), with Texas A&M Energy Institute leading the Gulf Coast Regional Manufacturing Center. The goals of CESMII are to (1) demonstrate at least 15% improvement in energy efficiency in real plant demonstration projects within five years, supporting a later goal of 50% improvement in energy productivity in 10 years; and (2) reduce the cost of deploying Smart Manufacturing initiatives to processing plant by 50% through the development tools and technologies. These includes developing platforms that enables machine-to-plant-to-enterprise real time sensing, instrumentation, monitoring, optimization, and control of energy. The idea of Smart Manufacturing integrates our expertise in high fidelity or advanced modeling of industrial operation, big data analytics, real time optimization and leverages high performance computing to achieve a seamless development of optimal processes and energy reduction.

The most recent advances in sampling technology and sensor development has enabled continuous online monitoring of polymerization reactions, leading to an opportunity for the application of advanced controllers. In this work, we develop model based advanced control strategies for batch polyolefin reactions via the PARametric Optimization and Control (PAROC) framework. Based on accurate linear representations of an initial high fidelity model, we formulate a Model Predictive Control (MPC) scheme that delivers a trajectory tracking problem. We solve the multi-parametric counterpart of the MPC problem (mpMPC) to determine the explicit expressions of the optimal control actions. Hence, the online computational burden of the optimization problem is reduced to a simple lookup table and affine function evaluation problem. Such an alleviation in the computational cost and ease of application allows for an opportunity for the integration of the control strategies in the Smart Manufacturing cloud platform to be used by industrial partners.

513861 Smart Manufacturing: A Must to Remain Competitive

Abstract:

Rapidly evolving technologies and products are posing enormous challenges and opportunities for chemical manufacturers. Meeting these challenges is a bare minimum for surviving in todayâ€™s dynamic global markets. Whereas to be competitive a manufacturer must be able to transform each and every rising opportunity into cash or strategic benefits.

Smart Manufacturing offers a multitude of capabilities to competitively deal with the ever increasing challenges and opportunities of chemical manufacturing.

Smart Manufacturing is a set of enabling technologies, methodologies and tools that can be deployed to improve the core business specific technologies, operations and work processes

The components of Smart Manufacturing are generic and are usually applicable to a number of different manufacturing processes.

In this presentation we will elaborate on main components of Smart Manufacturing in the context of chemical manufacturing. Proper deployment of tools such as big data analytics, process control and optimization, real-time fault detection & diagnosis, and predictive maintenance will also be described. A strategy to adopt Smart manufacturing will also be suggested.

513097 Development of an Advanced Control Philosophy for Dry Gas Combined Steam Reforming, to Operate the Plant Efficiently during Feed Composition Changes

Abstract:

Steam reforming of natural gas is one of the very important processes in the world now, as hydrogen, a key product of the reforming, continues to be in high demand for various petrochemical and chemical industries. As the gas reserves are depleting in the world, existing reforming plants are open to use natural gas from different sources. Different natural gas sources offer different compositions in the gas. This wide range of feed natural gas composition has significant effect on the performance of the steam reformers. In order to operate the reforming process at optimum conditions, precise control of reformer temperature, and steam intake are very crucial. Disturbances in the feed gas create instability in the reforming process. Furthermore, some steam reformers are operated with carbon dioxide added to the reformer feed, to have a lower hydrogen to carbon mole ratio, as required for downstream methanol or plasticizer alcohols processes. These dry gas combined steam reformers are operated at higher temperatures and lower steam to carbon ratios than that of pure steam reformers. Hence, frequent feed composition disturbances imply greater in-stability in the dry gas combined steam reformers in terms of reformer performance, reformer temperature, methane slip, fuel gas consumption and excess air in reformer box.

The existing control philosophy of the reformers cannot cope with the feed disturbances and can be effected by big swings in the reformer temperature, methane slip, fuel gas and steam & carbon dioxide flows in case of dry gas combined steam reformer. These disturbances may also have adverse effects on the plant efficiency, capacity, safety and reliability, and the plant integrity. In order to have a stable operation and cope with these disturbances, most of the existing plants are opting for advanced process control possibilities like Aspen DMC+ etc

In the paper, a new feed forward control strategy is described that can control the dry gas combined steam reforming process efficiently and effectively for a large composition variations. The basic methodology of the new control strategy is to control the steam reforming key parameters like reformer temperature, stream flow, process air, CO_2 flow, and Fuel gas flow, based on the carbon count of the natural gas in a feed forward way. Additionally, a small correction to the individual manipulators is done in a feedback control philosophy. This feed-forward & feedback control philosophy enables the operating plant to react to the disturbances quickly and efficiently. Hence, the plant performance disorder is kept minimum and huge swings in the parameters are avoided during the feed disturbances. Also, with this new control strategy, the variations in the plant parameters are reduced and stable and steady production of the hydrogen can save the energy about 0.5 GJ/ton of H_2 produced from the reforming process. More than the energy, the process operates more steadily and downstream processes that depend on the reforming process operate steadily as well. This developed strategy is tested and implemented and has improved the plant reliability and integrity.

514266 cMIST Compact Inline Gas Treating Technology

Abstract:

cMIST™: Compact Inline Gas Treating Technology

Shwetha Ramkumar, Norman K. Yeh, Don P. Shatto, Chuck J. Mart

ExxonMobil Upstream Research Company

Background

The natural gas industry is increasingly pursuing compact and low weight processing technology to meet the technological and economic demands of offshore, onshore, remote, and challenged gas processing. ExxonMobil Upstream Research Company has developed the Compact Mass transfer and Inline Separation Technology (cMIST™) gas treating system, a novel, compact and low weight processing technology platform, to achieve process intensification in processing facilities.

Aims and Methods

The cMIST™ system can replace conventional absorbers with in-line equipment for mass transfer and separation and can be applied to gas dehydration, selective H_2S removal and low level CO_2 removal. The first application of the technology, which is dehydration of natural gas to pipeline specification, has been qualified and licensed to Sulzer Chemtech for deployment on offshore and onshore facilities. It was demonstrated using a skid-mounted test unit in the Woodford trend area of Oklahoma, USA at an XTO production facility.

The selective H2S removal application is currently in demonstration phase at an XTO gas plant in Teague, TX.

Results and Conclusions

The cMIST™ dehydration system was tested to examine the dehydration capacity, operability, and reliability and it achieved pipeline specification over a range of process conditions and during sustained operation. The cMIST™ system for selective H_2S removal is currently being demonstrated with MDEA solvent and is achieving pipeline specification with significantly lower CO_2 pickup compared to a conventional contactor over a range of process conditions.

The cMIST™ system enables 50% system weight reduction and 50 - 80% absorber footprint reduction along with significant capital cost savings. The cMIST selective H2S removal system has added benefits of lower solvent circulation rate and the potential for elimination of the Acid Gas Enrichment (AGE) unit. The technology is highly modular to allow for simple transportation into remote, challenged, or offshore environments and provide installation configuration flexibility. The technology is also beneficial for debottlenecking and capacity enhancement applications.

513675 Novel Cryogenic CO_2 Removal- Operating Conditions and Economic Analysis

Abstract:

Natural Gas (NG) pretreatment systems have been simulated to remove more than 99% of the CO_2 from NG. Steady-state simulations with Aspen Plus^{TM} illustrate the effects of different parameters on CO_2 capture percentage. Vapor mole fractions of CO_2 and methane also have been investigated by changing the pressure at constant temperature. Increasing pressure increases CO_2 capture and hence methane mole fraction in the vapor. Consequently, mole fraction of CO_2 in vapor decreases. As pressure increases, the methane in the vapor also condenses, creating more liquid in which solid CO_2 can dissolve. Increasing methane increases both CO_2 and C_1 in the vapor stream and decreases the CO_2 capture percentage. Increasing ethane and propane increases both CO_2 and C_1 in the vapor and decreases the CO_2 capture percentage. An experimental bench-scale apparatus verified the simulation results. Exergy loss and heat exchange efficiency have been calculated and equipment sizes and cost estimates are complete. The results confirm the low exergy losses of 39.6 kWh/ kg treated NG for simulation and 44.8 kWh/ kg treated NG for bench-scale apparatus. High heat exchanger efficiency of +90% has been achieved. High pressure and low methane in NG composition results in high CO_2 capture, though the capture efficiency reaches a maximum at intermediate pressure.

512429 Dynamic Intensification of Separation Processes

Abstract:

Process intensification (PI) has captured significant attention as a set of innovative design principles that can dramatically enhance process performance, reducing energy/capital cost and increasing process efficiency[1]. PI technologies usually integrate multiple coupled phenomena into one physical device to optimize mass, heat transfer and reduce equipment size. However, the advantages of PI, including reduced holdups, increased safety and more nimble operation come with some drawbacks and operational barriers. These include a reduction in the number of degrees of freedom (DOF) available for control, as well as a narrower window for optimization compared to conventional process equivalents[2].

In this paper, we will introduce a new dynamic operation paradigm that can be applied to existing intensified equipment to further enhance process performance. In particular, we study the use of periodic operation to manipulate and hijack the competing phenomena present in an intensified system to the benefit of economic indicators such as energy efficiency and yield. We refer to this approach as â€œdynamic process intensification.â€

Motivated by their large energy consumption across the chemical and petrochemical industries, we focus specifically on the dynamic intensification of distillation systems. Using the canonical model of a binary distillation column, we demonstrate theoretically the benefits of dynamic intensification. We then carry out extensive dynamic optimization and simulation studies to confirm the results of the theoretical findings and quantify the magnitude of the economic benefits available in a system of practical relevance.

References

[1] Creative Energy. European Roadmap for Process Intensification. 2007

[2] Schembecker, G.; Tlatlik, S. Process Synthesis for Reactive Separations. Chem. Eng. Process. 2003, 42, 179-189.

Top Dividing Wall Columns in Process Intensification

Manish Bhargava, Roomi Kalita

GTC Technology US LLC., Houston, USA
mbhargava@gtctech.com
rkalita@gtctech.com

Dividing Wall Column (DWC) technology provides refiners with a unique way of improving the efficiency and profitability of the traditional refining techniques. This technology, which has been in use since the 1940s, can be modified for various applications ranging from naphtha/reformate splitters to gas plants and can even be used to transform established technologies like isomerization and naphtha hydrotreating.

In this paper, we discuss the distinct advantages that DWCs can offer to improve and modify traditional gas plant process schemes with emphasis on top dividing wall columns. DWCs are often characterized by the presence of a vertical wall in a regular distillation column. When the distinctive wall is present at the top of the column, the column can be referred to a top dividing wall column.

Top DWCs offer a host of advantages, especially in gas plants which are generally operated under high pressures. Unlike a conventional distillation column, the top of these columns is segregated into two separation zones within the same shell with no intermixing between the two zones. As a result, they can be operated in separate modes and basically behave as two independent columns. Absorption and distillation can be carried out in a single column. This principle forms the basis of GT-LPG MAX® technology.

The concept of Top DWCs can also be implemented in a reformer unit gas plant and a FCC (fluid catalytic cracking) unit gas plant, wherein two or more columns operate in a sequence and under similar operating pressures and temperatures. The DWC will potentially combine the operation of two or more columns, thereby minimizing both capital and energy consumption of the whole sequence.

Top Dividing Wall Columns in LPG Recovery from Off gas

LPG (Liquefied Petroleum Gas) recovery units are an integral part of any refinery. In general, in LPG unit two or more columns remove the C_1-C_2 components along with other non-condensable gases, while concentrating the C_3-C_4 components to generate LPG. Figure 1 shows a standard design consisting of a deethanizer column followed by a depropanizer column. The deethanizer column operates at a high pressure, wherein the C_1-C_2 components are removed as the top product. The C_3 rich (or C_3-C_4) stream is removed in the next column. The recovery of LPG in this design is generally poor due to the high operating pressures and propane losses in the off-gas. In some processes, these losses are mitigated by using refrigeration in the overhead system. This increases the capital cost of the process.

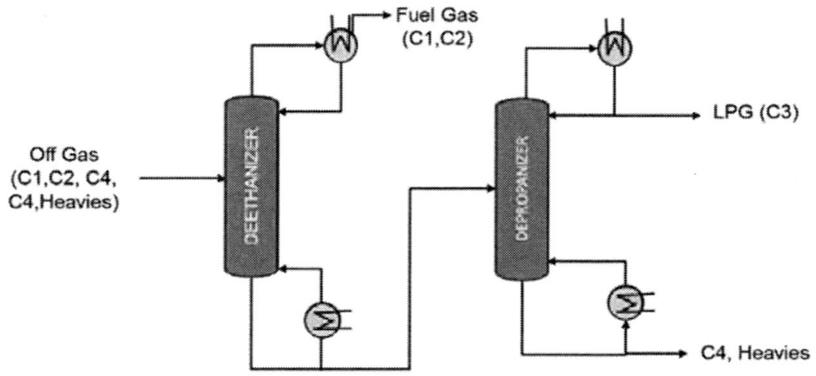

Figure 1: Conventional LPG Recovery Unit

These problems can be rectified by using a top dividing wall column. Doing so allows the top section of the column to be operated under two different unit operations, i.e., absorption and distillation. The section of the column where the feed enters operates under reboiler absorption. The other section of the column separates a light liquid product and a heavy bottom product using regular distillation. This arrangement offers the following advantages.

First, use of absorption allows the column to be operated at a lower pressure. The absorbing solvent captures the valuable C_3-C_4 components in the off-gas and moves them towards the bottom of the column. These components are then concentrated on the other side of the DWC using distillation. For feeds which contain a good amount of C_5 and heavier components, a portion of the heavy bottom product can be used as the absorption medium (Figure 2a). Using the bottoms is highly beneficial when the feed contains at least 30-40 wt% of C_5s. Table 1 gives a sample GT-LPG MAX® feed (Bhargava et al., 2015). If the C_5 concentration in the feed is low, the bottoms solvent can be supplemented by an additional heavy oil solvent from nearby processes.

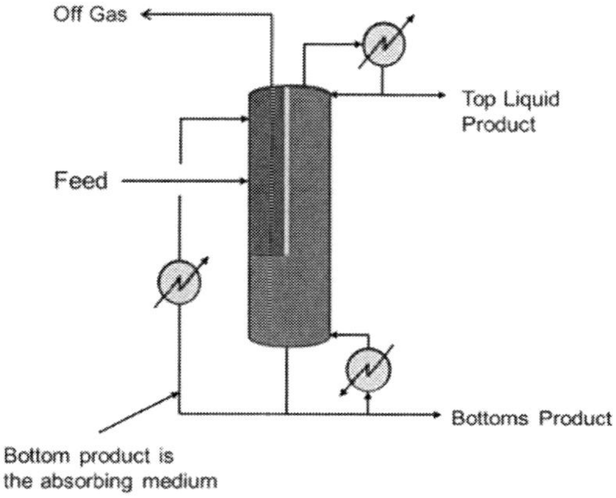

Figure 2a: GT-LPG Max® process scheme; Figure 2b: Inside view of a top dividing wall column

Table 1: Sample Feed Composition for GT-LPG MAX® Design

Product Class	Feed Composition		Product Specifications
	Components	wt.%	
Fuel Gas	Hydrogen	0.0	Minimize Propane
	Methane	0.3	
	Ethylene	0.0	
	Ethane	3.0	
	H_2S	0.5	
	CO_2	0.0	
LPG Product	Propene	0.1	Propane > 90.0 wt.%
	Propane	11.7	Ethane <1.0 wt.%
i-C_4 Product	I-Butane	8.3	Propane < 3.0 wt.%
			n-C_4 < 4.0 wt.%
n-C_4 Product	n-Butane	19.6	i-C_5 < 3.0 wt.%
			i-C_4 < 5.3 wt.%
Debutanizer Bottoms	C_5 and heavier	56.6	

Due to the low column operating pressure, water-cooled partial condensation is used in the off-gas side of the column. This is highly economical as compared to using refigeration to mitigate C_3-C_4 losses.

Additionally, this DWC design sees a reduction in energy consumption by about 10-30% as compared to traditional columns with better product recoveries (Table 2).

Table 2: Project Economics of GT-LPG MAX® Design

Variables	Units	Conventional Design		GT-LPG MAX® Design
		Deethanizer	Depropanizer	
Operating Pressure	MPaG	3.1	2.1	2.1
Reboiler Duty	MMKcal/hr	5.5		5.0
Propane Recovery	%	55.0		97.0

Top Dividing Wall Columns in Reformer Unit Gas Plants

A typical reformer unit gas plant usually has a depentanizer column, followed by a debutanizer and a deethanizer. In Figure 3, the depentanizer column removes the C_5- cut from the reformer bottoms as the top product. C_6 and heavier cuts are removed as the bottom product. In the next column, the C_5s are concentrated in the bottom of the column. The C_4 and lighter cut is then processed in a deethanizer column, which removes a LPG cut at the bottom.

The three columns in this sequence operate at high pressures. Additionally, the debutanizer and deethanizer columns see partial overhead condensation. Due to these, a major portion of the LPG components (C_3-C_4) end up in the off-gas. These components can be recovered by use of a dividing wall column configuration.

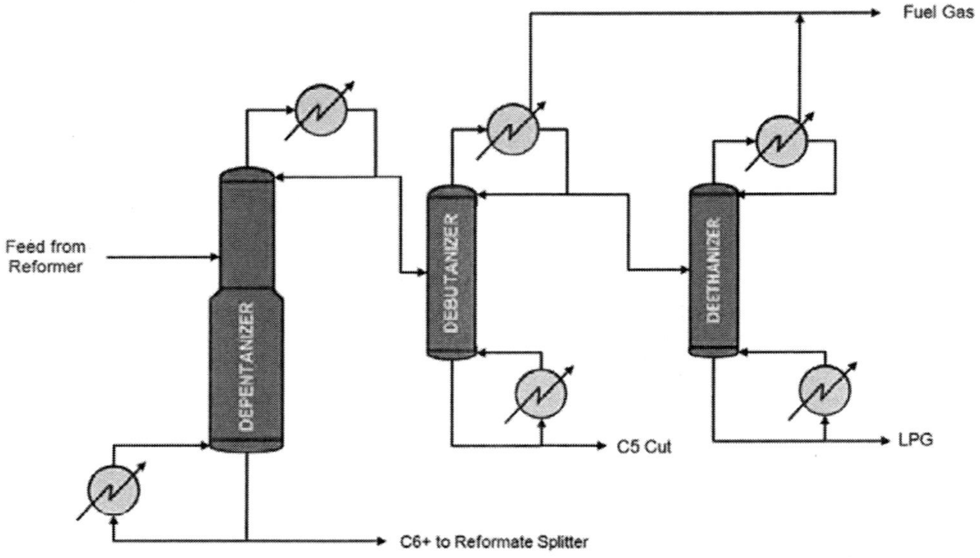

Figure 3: Conventional Reformer Unit Gas Plant Configuration

As shown in Figure 4, the operation of the deethanizer can be squeezed into the bigger depentanizer column by means of a dividing wall. The column has two distinct zones within, a pre-fractionation zone and a main fractionation zone. The pre-fractionation zone sees a combination of absorption and distillation. The lights are separated on this side as the off-gas, with the absorption medium reducing the liquid losses. On the main fractionation side, the C_5 and lighter components are removed as the other top product. Similar to a GT-LPG MAX® design, a portion of the heavy bottoms (mainly C_6-C_7) is used to strip the gas. The GT-DWC® depentanizer column is operated at a lower pressure as compared to the conventional column.

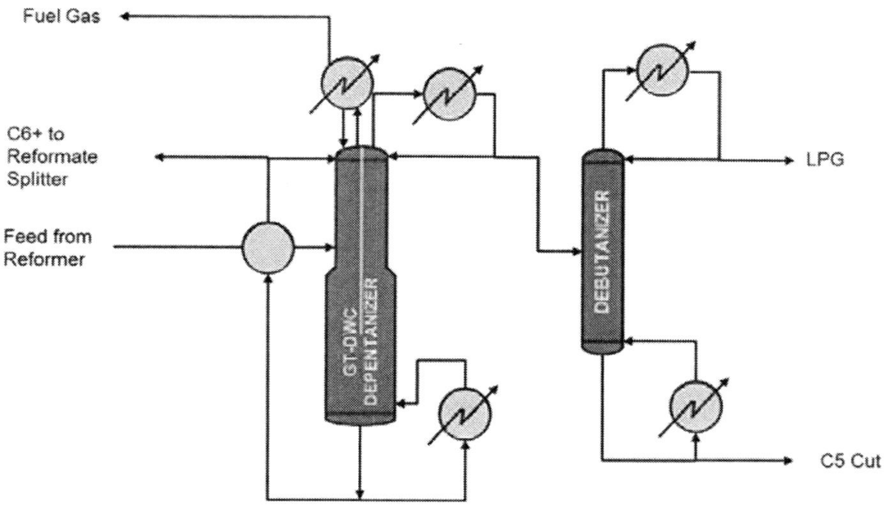

Figure 4: Reformer Gas Plant Configuration with GT-DWC® Technology

Since the majority of the lighter components is removed in the first column, the subsequent debutanizer column operates with a total overhead condenser. This column removes the LPG cut and the C_4 cut. Table 3 compares the performance of the two configurations. For the same heating duties, the DWC configuration generates better product recovery while minimizing the amount of equipment needed. The DWC configuration has a lower total installed cost over the conventional configuration.

Table 3: Performance of GT-DWC® vs Conventional Design

Parameters	Units	Conventional Design			GT-DWC® Design	
Columns		Depentanizer	Debutanizer	Deethanizer	Depentanizer	Debutanizer
No. of Trays	-	40	33	27	40	40
Operating Pressure	MPaG	0.9	1.1	3.1	0.5	1.1
Total Heating Duty	MMkcal/hr	37.1	2.2	0.5	29.0	2.7
Total Installed Cost	MMUSD	10.7			8.5	
Feed	tonnes/hr	386.8			386.8	
C6+ Cut	tonnes/hr	365.5			365.5	
C5- Cut	tonnes/hr	8.1			8.2	
Off Gas	tonnes/hr	5.8			1.7	
LPG	tonnes/hr	7.4			11.4	

Top Dividing Wall Columns in FCC Unit Gas Plants

Figure 5 shows a traditional FCC unit gas plant. The process scheme primarily consists of absorber/stripper columns, where the gases and the liquid in the feed are separated. Unstabilized naphtha (supplemented with a portion of stabilized naphtha bottoms) prevents the C_3-C_4 losses from the absorber column. The remaining C_3-C_4 are extracted by using lean oil in the sponge absorber column. Sour fuel gas and rich oil get separated at the top and bottom. The liquid product from the stripper column is processed in a debutanizer column. Sour LPG is extracted as the top product, while stabilized naphtha is recovered at the bottom.

The conventional design consists of three stages for the separation, the absorbing and stripping section followed by distillation in the main debutanizer column. Using DWC technology, the three stages can be integrated into a single column (Figure 6).

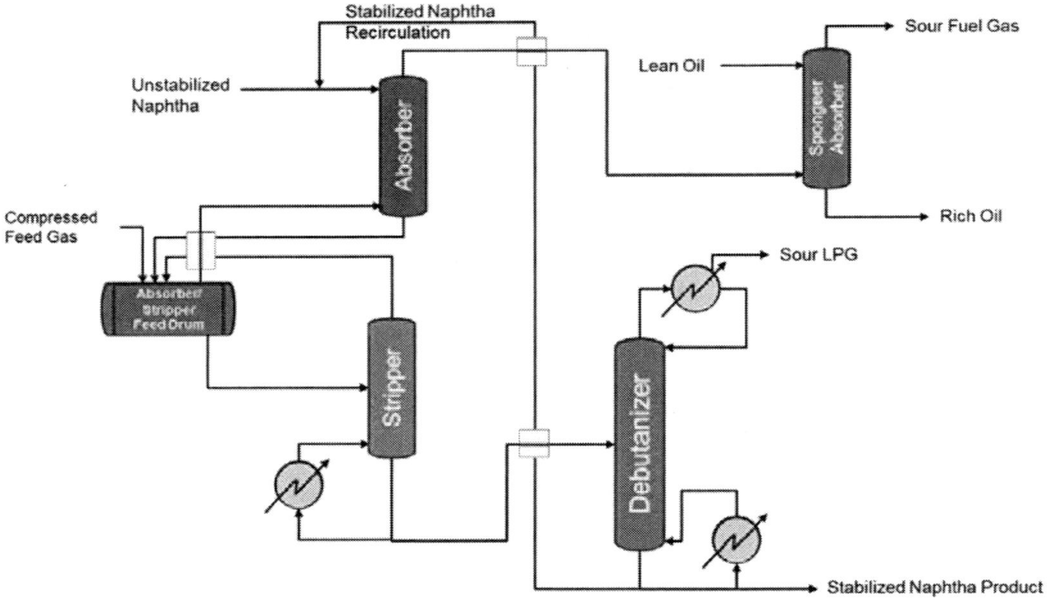

Figure 5: Typical FCC Unit Gas Plant Configuration

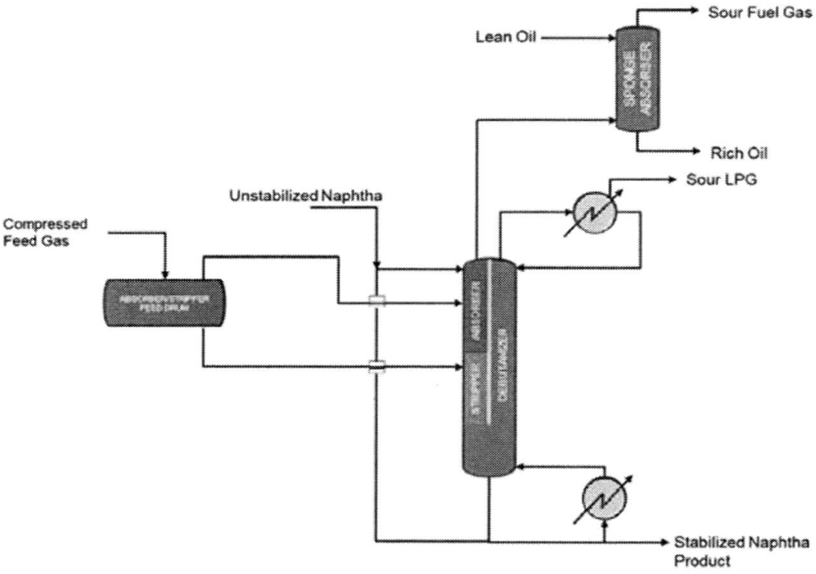

Figure 6: FCC Unit Gas Plant Configuration with GT-DWC® Design

Due to the segregation of the top halves, the two sides of the column behave as individual columns with independent overhead systems and controls. The parallel zones provide similar stages for separation in a single column. A portion of the bottoms product is circulated as the absorption medium in the section where the feed enters. This operation is similar to that of the absorber in the conventional design. The liquid portion of the feed enters below the absorption section. The C_3-C_4 components move down the wall aided by absorption. Once across the wall, these components are concentrated as the top product by rectification and stripping. The stabilized naphtha is removed as the bottom product. Table 4 gives a comparison between the two designs.

Table 4: Comparison between Conventional Design and GT-DWC® Design

Parameters	Units	Conventional Design	GT-DWC® Design
No. of Columns	-	4	2
Total Heating Duty	MMKcal/hr	9.9	7.9
Utility Savings	%	-	20.0
Feed rate	tonnes/hr	55.3	55.3
LPG Recovery	Vol %	85.1	85.1
C4 Recovery	Vol %	98.8	99.6

Conclusion

Gas Plants are a vital component of any refinery. These columns have always been operated using well-established distillation techniques. However, using the Dividing Wall Column concept, primarily top DWCs can generate a leaner and efficient system. DWCs not only lower the equipment count (columns, pumps, and reboilers' et al.), these columns also lower the energy consumption of any configuration. On occasions, these novel distillation columns can even provide better and sometimes new products as compared to the conventional columns. Integrating DWCs in gas plants can virtually change the way the refining industry improves the profitability of their gas plants in the coming years.

References

Bhargava M., Nelson C., Gentry J.C., Siddamshetti, V., 2015, Improved Distillation Efficiency, Hydrocarbon Processing (January)

512482 A Kinetic Analysis of the Shipp Synthesis of 2,2',4,4',6,6'-Hexanitrostilbene from 2,4,6-Trinitrotoluene

Abstract:

2,2â€™,4,4â€™,6,6â€™-Hexanitrostilbene (HNS) is an important explosive utilized for multiple purposes due to its thermally stable nature and insensitivity to impact and electrostatic shock. The most common synthesis route used for manufacturing HNS is known as Shipp synthesis and involves using an alkaline sodium hypochlorite solution to facilitate the conversion of 2,4,6-trinitrotoluene (TNT) to HNS. The goal of this study is to develop a reaction network and the corresponding kinetic parameters for the Shipp synthesis method. This reaction network can then be utilized to optimize the reaction conditions and show how the production of HNS using Shipp synthesis could be converted from a batch process to a continuous one. A key issue surrounding this synthesis route involves the low solubility of both the reactants and products, therefore large importance is placed on the solubility modeling of the reacting system.

512964 Halloysite Nanotube Nanocomposite: Processing and Characterizations

Abstract:

Nano-clays have gained increasing popularity in practical applications including functional polymer-clay nanocomposites (PCNâ€™s), material compatibliziers, polymer and rubber processing aids, performance enhancing nanofillers, rheological property modifiers, thermomechanical property enhancers, use in controlled drug release agents, and much more. Among the members that are classifiable under nanoclays, halloysite nanotube (HNT) is of particular interest, based on its abundant availability as high-purity naturally-occurring minerals, unique tubular nanostructure, high aspect ratio, inherently rich functionality, a nearly uniform ratio (2:1) of silica to alumina, excellent cation exchangeability, retainability of both hydrophilic and hydrophobic chemicals, and more. In this paper, fabrication, processing, and characterization of HNT-nanocomposites will be examined and a preferred approach will be presented. The relationship between the thermomechanical and rheological properties and the specifically enhanced functions will also be investigated on the prepared HNT-nanocomposites.

512987 The Development of a Novel Method Using Cryogenic Grinding to Produce Chitin Nanofibers for Blending with Biodegradable Polymers

Abstract:

As plastics continue to buildup in our landfills, the demand for biodegradable polymers is ever increasing as an alternative to conventional polymers, which may take hundreds of years to break down in the natural environment. However, many of the current commercially available biodegradable polymers lack all of the desirable properties of conventional thermoplastic polymers. One such way to improve existing biodegradable polymers in functional applications is by reinforcing them with nanofibers in forms of nanocomposites. Of more recent interest in this area, chitin nanofibers have shown to increase Youngâ€™s modulus and toughness when incorporated into thermoplastic polymers. Chitin is a highly abundant natural biopolymer which is second only to cellulose. Chitin is the primary component of the exoskeletons of most arthropods, including crabs, prawns, and insects. It can also be found in fungi cell walls. Unlike cellulose though, the majority of chitin is just discarded as commercial waste. The exoskeletons are largely made up of a chitin fiber matrix along with minerals and proteins. Once the minerals and proteins are removed, the chitin fibers are strongly held together by hydrogen bonding which can make fibrillation quite difficult. Chitin is also not soluble in most common solvents. There are currently various methods in the literature used to produce chitin nanofibers including electrospinning, grinding with specially designed grinding stones, and high-pressure homogenization, but currently no methods that have been commercialized. Additional research is needed to develop new methods and improve existing methods of chitin fibrillation. In this study, a cryogenic grinder was used to grind commercially purchased chitin with additional modifications made to the grinding process to improve fibrillation. Results were analyzed by SEM, TGA, DSC, and XRD.

513484 Young Managers with Experienced Employees

Abstract:

There may be times throughout the career of a young engineer in which they will be placed in a role in which their colleagues - who have more experience - are reporting to them. This session will touch on the issues that can arise from this situation and what we can do (as young managers) to steer and control the relationship in order to maximize productivity.

Brandon Harding is currently a Chemical Process Engineer at the Holston Army Ammunition Plant (operated by BAE Systems) located in Kingsport, TN. In just 2 short years with BAE Systems, Brandon has been able to successfully implement several projects that save millions of dollars annually and has been awarded multiple Business Leader & Chairman Awards. As a process engineer, there are times when you do not directly manage colleagues - particularly those with more experience - but will require them to work for you in order to drive change. Brandon will discuss some personal experiences in which he had do "manage" those with more experience in order to accomplish a task or complete a project. Specifically, we will discuss how each issue began and what was done to control the situation and direct all project members toward a more productive environment. This will be an interactive portion (with the audience) of the session.

510755 The Sartomer Success System (3S): Achieving Operational Excellence

Abstract:

The Sartomer Success System (3S) was initiated at the Sartomer Chatham facility in June of 2015 and, based on that success, deployed at the Sartomer West Chester facility in June, 2017. The 3S Process is a site-wide Operating Excellence Initiative, creating a culture that allows our most valuable asset, our employees, to directly contribute to the success of the site. The 3S Process is used to identify and create employee-driven Pillar Teams, each with specific elements to address. These Pillar Teams are site specific, identified by the Plant Leadership Team, to address specific needs of the site aligned with the Company and Site Vision. For West Chester, these teams include: Process Safety, Daily Management Systems, 5S, Training, Focused Improvement, and Planned Maintenance, in addition to the 3S Steering Team. The 3S Steering Team is comprised of each of the Pillar Team Leaders, Plant Management Staff, and the Plant Manager. To date, more than 90% of our employees at Chatham (75% at West Chester) have voluntarily participated on one or more Pillar Teams.

Prior to 3S Deployment, Chatham struggled to engage hourly employees in improving the site's performance. Morale at the site was poor and product quality suffered. Chatham had experienced a number or process safety incidents, with poor self-reporting of smaller incidents (including near miss incidents), and had systemic issues with a number of other process safety (PSM) elements. Communication, which includes communication between management and the hourly associates, was also cited as an opportunity in employee surveys. For West Chester, similar gaps existed, with a greater need to improve onboarding of new employees due to a large number of employee retirements.

Since deploying 3S, Chatham has achieved "best ever" levels of personnel and process safety, product quality, cost control, and innovation. Similar gains are expected for West Chester. From a safety perspective, Chatham recently met the milestone of Sartomer's "double zero", which is more than 1 year without a major process safety incident (loss of primary containment or environmental release) or an OSHA recordable injury. Chatham has not had an OSHA recordable in more than 4 years, and has operated more than 11 years since it's last lost time incident. West Chester also reached more than 1,000,000 man hours since its last lost time injury earlier this year and is preparing for certification by OSHA's voluntary protection program (VPP) in early 2018. A business objective for Sartomer to achieve a "double zero" across the global business (4 manufacturing sites) in 2018.

The 3S Process is being used as a benchmark for Arkema, with many of the 3S elements being incorporated in to Arkema's own global Operational Excellence Initiative, SMART. Chatham was selected as a pilot facility for SMART in the Americas in early 2016, with the Plant Manager of Chatham joining the SMART design team mid-year, participating in the EU introduction of SMART at the EU Plant Manager's meeting, and assisting and creating the program for the SMART introduction at the America's Plant Manager meeting in late October, 2016. West Chester is deploying SMART concurrently with 3S starting in October, 2017.

Each Pillar Team's progress is displayed prominently on Pillar Boards, which includes: team membership (with pictures), the team's mission statement and objectives, key performance indicators (KPI's) to track monthly progress against these objectives, minutes, action items, and path forward. The Pillar Boards are maintained by the respective Pillar Team. This is to ensure visibility and accountability for all Pillar Teams, as well as encouraging further employee participation.

VSP, a Sartomer manufacturing site in France, also volunteered to be a SMART Pilot facility at the EU Plant Manager's meeting. The 3S Process is also planned for expansion across Sartomer at the Nansha (China) facilities. In short, all Sartomer facilities are currently building on processes learned with 3S, elements of which are being expanded across Arkema with the global deployment of SMART.

514228 1st Year Experience of a Young Manager in a Chemical Manufacturing Unit

Abstract:

Donna Bryant has been Unit Superintendent of the Environmental Operations Unit at Syngenta in St. Gabriel, LA for a little over a year. In this time she has learned a lot about managing people, especially people that are older than her. In this presentation, she will be discussing what she has learned over this past year, what she wishes she had known when she started, and what skills she plans to expand upon.

513716 Justifying Energy Savings in Ethylene Crackers in a Low Energy Price Environment

Abstract:

Global energy prices have dropped to their lowest level in a decade and have taken with them the days of easy justification for energy improvements. In specific countries where energy prices are still subsidised, this is a familiar situation in everyday energy management. The subject of this paper is the different ways that have been used to improve the economics of energy projects in low price situations in order to make viable project investments.

Case studies will be used to illustrate such methods as:

- Debottlenecking energy critical equipment to increase throughput,
- Overcoming an energy cap to improve severity and product yield,
- Making energy savings to meet legislative targets, and
- Using energy reduction to justify changing catalysts to improve yield.

511653 Retrospective Health Impact Assessment for Pm 2.5 Pollution in Mexico City 1990-2015

Abstract:

Fine particulate matter has a well-documented association to mortality. We have conducted a retrospective health impact assessment to quantify the lives saved in Mexico City from 1990-2015 due to PM 2.5 reduction resulting from pollution control policies, utilizing the Integrated Exposure Response model developed for the Global Burden of Disease study. In addition, we monetized the health benefits using city-specific value of statistical life (VSL). Our analysis suggests that 21,000 lives saved during that period and benefits in the order of 18 billion 2015 USD. We also developed a novel model to retrospectively decompose the health benefits and attribute them to one policy utilizing sulfur reduction in fuel policy as an example. Our model indicates that this policy has resulted in 1,600 lives saved. This suggests that Mexico City air pollution policies could serve as a model for developing countries and the industry with environmental regulations at infancy stage.

512814 Sustainability Analytics - the Critical Link between Sustainability and Business Strategy for Actionable Insights

Abstract:

Advances in Enterprise Systems Are Making It Feasible for Corporations to Track and Transform Sustainability Performance. the Materiality of These Seemingly Non-Economic Impacts Is the Critical Link between Sustainability and Business Strategy. Business Analytics Enable Organizations to Convert Raw Data into Actionable Insights to Achieve Their Sustainability Goals. Sustainability Analytics Analyze the Data and Provide Insights That Help Transform Sustainability Information into Action across the Value Chain and Life Cycle. This paper will Provide Business Leaders with Insight into How to Determine Which Sustainability Metrics Are Material to Them and Relevant to Their Business, and How to Incorporate Them into Corporate Strategy, Resource Allocation Prioritization and Decision Making Process.

Based on My Book Introduction to Sustainability Analytics Under Publication By CRC Press

512839 The Reaction Mechanism and Kinetics of Magnesium Oxide for CO_2 Immobilization

Abstract:

The CO_2 immobilization is significant for the sustainable energy and chemistry industries. And the magnesium oxide is a clean and economic agent using for the CO_2 immobilization. Therefore, the reaction of CO_2 with magnesium oxide suspended in water was investigated, the mechanism proposed, and the kinetics regressed as well. The CO_2 pressures from 1 to 5 (MPa), as well as ratios of solid to liquid, stirring speeds and temperatures, significantly influenced the dissolution of magnesium oxide. And as high as 26.63 g/L MgO was dissolved in the solution under the condition of CO_2 pressure 5 MPa at 50 °C.

In the process, the magnesium oxide became magnesium hydroxide through hydration reaction, and then the hydroxide dissolved by carbonated water. In the carried out reaction of CO_2 with magnesium hydroxide suspension, no solid product was found covering the solid $Mg(OH)_2$. Therefore, a reaction mechanism based on the hydration of Mg^{2+} was proposed, where the concept of â€˜progressive liquid filmâ€™ was introduced to the reaction system. And the derived kinetic model for the dissolution was fitted the experiment data well. The pressurized magnesium hydroxide dissolution was found to be rate-controlled by the diffusion in the progressive liquid film around the solid material. The apparent activation energy, obtained from 17.1 to 25.4 (kJ/mol) in the experiments, indicated also that the dissolution inclined to be diffusion rate controlled. Moreover, the apparent diffusion coefficients of H^+ in the liquid film were calculated and an empirical expression was obtained for the first time. The kinetic model might be applicable to other dissolution reactions similarly with just soluble product.

513495 Smart Systems for the Process Industries

Abstract:

ABSTRACT for AIChE 2018 Spring Meeting

Smart Systems for the Process Industries

Samvith Rao and James C Cross III

The MathWorks, Inc.

samvith.rao@mathworks.com

508-647-4487

An unprecedented and increasing level of device and computing capability and connectedness is intensifying competition and paving the way for new business opportunities. The challenge for existing companies and new ventures alike is to: determine what information is important; decide how to collect and process it; and conceive, develop, and offer new products and services that deliver value from it. This is the aspiration of the Industrial Internet of Things (IIoT), and Industry 4.0.

Specifically, the proliferation of inexpensive sensors, wireless communications, big data and cloud infrastructure, machine and deep learning algorithms, and ever-faster high-performance computing platforms is ushering in a new era of intelligent and autonomous systems, including self-driving cars, smart industrial systems, and automated warehouses. Moreover, the confluence of these technologies has given rise to a new type of industrial asset – a virtual, real-time replica of a companion physical system, the so-called *digital twin* – that can be leveraged for a wide range of operational benefits, including process optimization, predictive maintenance, fault detection, model-based control, and dynamic pricing.

In this talk we describe the individual enabling technologies, present examples of how they are being developed or used commercially, and offer perspectives on their adoption and potential within the process sector.

512516 Smart Damper Design Provides Better Control of Fired Heaters

Abstract:

Fired heaters are major consumers of energy in the Refining and Petrochemical Industries. In most of the natural draft fired heaters, one of the important parameters that is controlled by the operators is the furnace draft using stack dampers. Historically stack damper quality in the industry is very poor. The current design of the stack dampers is inadequate to control the draft effectively. As a result most of the stack dampers are left fully open in the industry. The average draft in the fired heaters is maintained at almost 3-4 times the value recommended causing a lot of tramp air in the system.

In this paper; we are going to talk about how our patent pending Smart Stack Dampers can ensure proper draft control in fired heaters. We have made several improvements in the design and construction of the dampers. Smart dampers are designed to be very reliable and operator friendly. They provide a much better control of the draft. Smart dampers can be configured for automatic or manual draft control of the heater

513860 Application Centralized Information Technology in Process Design Package Development

Abstract:

Smart Manufacturing solution is built on information technology, researchersâ€™ ingenuity to create innovative product, continuously analyzing market and manufacturing data to discover the market needs and signal problems impacting your costs, quality, and production of the existing technology. Centralized database technology with advanced simulation are keys to quickly delivery of the new products reliably, safely, and environmental friendly.

This presentation provide an introduction for applying centralized information technology in process design package development (1) evaluating new innovative technology for its reliability, product profitability, global market potential, sustainability and technology; (2) Aligning proven and conventional technology to form an integrated commercial process based on commercial database; (3) develop the process design package and expand the process globally in a short timeframe.

512585 Use of Dividing Wall Technology to Increase Throughput and Improve Product Specifications in Two-Diameter Columns

Abstract:

Two-diameter section columns are quite common in the refining industry. Standard examples include naphtha splitter columns, stabilizer columns or depropanizer columns. These columns are encountered in applications wherein the process sees different vapor-liquid flow within the top and bottom section of the columns. In such applications, two-diameter section columns are favored by refiners over a single diameter column due to savings in fabrication of such columns. However, for any capacity increase or improving the process separation, this smaller section acts as a bottleneck due to the inability to handle increased vapor-liquid loads. Use of Dividing Wall Column (DWC) technology will allow the refiner to circumnavigate the problem. DWC allows the optimization of the lower diameter section to achieve better separation as well handle the increased capacity from the original design.. Two case studies based on actual installations will be presented:

- A conventional two-column fractionation sequence is revamped to a single two-diameter DWC column to produce the same product specifications as the original design. The second column in the sequence is idled following the revamp.
- An abandoned two diameter column is revamped for maximizing IC5 Recovery.

Authors:

- Manish Bhargava, mbhargava@gtctech.com
- Roomi Kalita, rkalita@gtctech.com

GTC Technology

900 Threadneedle St., Suite 800

Houston, TX 77079

281-597-4800

513423 Process Intensification of Distillation Using a Microwick Technology Platform

Abstract:

Process intensification of distillation has been advanced using a microwick technology platform that has been demonstrated for multiple gas-liquid processes, including partial condensation, absorption, stripping, and distillation. Mass transfer processes are intensified by constraining the liquid phase within thin wicking structures, which for distillation dramatically reduces the height of an equivalent theoretical plate (HETP). Multiple separations have been demonstrated with microchannel distillation (MCD) devices, including fractionation of JP-8, a military diesel fuel, purification of n-octane, separation of propane and propylene, and isotopic enrichment of methane. Results have yielded HETPs as low as 0.33 cm and up to 60 separation stages were accompished in a single device with an active length of 25 cm. Dimensional analysis of the data from the variety of separations provide insights into the performance and scaleability of the technology. Computational fluid dynamics models of the processes indicate the potential for an additional factor of 10 reduction in HETP. Comparisons are made to other microfluidic distillation technologies as well as to commercial advanced structured packings.

513707 Improving Plant Economics By Upgrading Low Value Hydrocarbon Stream By Enrichment of C_{9+} Compounds Using Divided Wall Distillation Columns

Abstract:

The shift towards lighter feedstockâ€™s for Poly Olefins production, due to lower feedstock price and higher profit margin, has resulted in lower production of heavier aromatic fractions and in particular the C_{9+} fractions. As a result, over the last 5 years the supply of C_{9+} fractions has not catch up with increase in demand. Further, intense competition between petrochemical complexes has forced companies to retrofit plants for comprehensive utilization of all streams in the complexes for better operating margin. Among other things that are considered for up-gradation, enrichment of heavier hydrocarbons from the Pygas stream has been pursued intensely. In this work, the process intensification in C_{9+} stream enrichment from Pygas using non-conventional divided wall column structures is examined. The study is a first of kind to demonstrate the separation of bulk chemicals using divided wall columns. Further more, the studies indicated that by replacing conventional column sequence with the divided wall column structure, for a given product quality, a savings in the separation stages by 18%, energy reduction by around 32% can be realized.

513890 Overcoming Foreseen and Unforeseen Problems with Oldershaw Column Designs in a Pilot Plant

Abstract:

During a pilot plant design, the team recognized a potential problem with standard Oldershaw tray designs handling unusual vapor / liquid ratios, which would be the norm in this particular system. Cold-flow testing with several off-the-shelf Oldershaw column sections demonstrated that the low vapor flow would likely result in poor tray loading and efficiency during operation. Although very little published data is available for the sizing of Oldershaw columns, the team successfully designed custom trays with a unique hole pattern to handle these unusual conditions.

However, upon startup, downcomer flooding problems were encountered in the lower portions of several tray sections. Testing and troubleshooting identified a simple but easily overlooked issue with the bottom tray outlet design that was leading to the inadequate performance.

Once corrected, the trays performed well with good loading. Tests using chemicals with well-understood VLE data validated the tray efficiency, and the results matched predictive correlations for efficiency and pressure drop quite well. This gave the team the confidence necessary to trust that the system could be used to validate the actual process VLE data and scale-up the system for commercial design.

This presentation will qualitatively cover the discussions that led to the identification of the vapor loading concern, the testing that validated it, and the hurdles overcome to produce a sound design basis for the tray design. It will cover the hydraulic problems encountered that led to the downcomer flooding and the corrective actions taken to fix that problem. Lastly, it will cover the testing used to validate the column performance so that the data collected during actual pilot plant operation could be trusted to scale up to a commercial plant design.

513876 Pilot Plant Success Stories from Xerox Research Centre of Canada

Abstract:

The Xerox Research Centre of Canada (XRCC) has been conducting advanced materials research, development, and scale-up since 1974. The XRCC Pilot Plant has been vital for scale up and derisking numerous chemical processes on the path to commercialization. Its success has been demonstrated by the fact that all Xerox printers on the market today contain at least one technology that was developed at XRCC. A few of these success stories will be discussed during this presentation.

513858 Piloting to Prevent Failure

Abstract:

Pilot plants are a key risk mitigation tool in commercial process development. If properly designed, they can become a test bed for the mitigation of risks related to the fundamental underlying process (reaction engineering, catalyst deactivation/regeneration, byproduct accumulation, recycle stream management etc.). They can also be used to resolve issues related to materials of construction, scale dependant operations such as filtration, and other parameters which are difficult or impossible to resolve solely with recourse to the literature or simulations.

In this presentation, a number of pilot plant projects across a diverse swath of the chemical process industry are examined, demonstrating how certain learnings from the pilot program design or execution mitigated key scale-up risks, identified new problems not previously anticipated, or identified a key roadblock to commercialization for a new process technology.

513229 The US Doe's Abpdu: Enabling Commercialization in the Bioeconomy through Rapid Process Prototyping and Piloting

Abstract:

The Advanced Biofuels and Bioproducts Process Development Unit (ABPDU) was established in 2009 and commissioned in 2012 at Lawrence Berkeley National Laboratory by the US Department of Energy's Bioenergy Technologies Office. The ABPDU's mission is to optimize and scale up biomass-based conversion technologies in collaboration with partners and collaborators from industry, other National Labs, and academic research institutions. Recently, the ABPDU's focus has included the production of not just biofuels like alcohols and hydrocarbons but also bio-based chemicals and materials such as commodities, specialties, biogels, and plastics; as well as proteins such as industrial enzymes and food and feed ingredients. To date the ABPDU, located in the heart of the San Francisco Bay Area, has partnered with more than 30 companies. This paper will focus on some of the success stories from our industrial collaborations, where our corporate partners have gone on to launch products, attract private capital investment, receive Federal funding, and develop technology transfer packages for larger scale demonstration and commercial production based on our work with them. A key element of the ABPDU's collaborations with our partners is the approach to rapidly, safely, and robustly execute novel processes from a variety of feedstocks using distinct and custom unit operation configurations. This includes optimizing not just the core fermentation or chemical process technologies, but also any required upstream and downstream separations, purification, and polishing steps. To achieve this the ABPDU performs process hazard reviews prior to project start and ensures team training and equipment readiness are up-to-date. Project management, analytical chemistry, scale-up / scale-down trials, and techno-economic modeling underpin these successes, and allow the team to rapidly develop and de-risk the next generation of bioproduct technologies.

513267 Integration of Computational Dynamics, Data Analysis Tools and Process Simulation for Modeling of Waste Water Treatment Plants

Abstract:

On-line and off line plant data, process flowsheet simulators and CFD are synergistically integrated for development of operating policies for waste water treatment plants. The waste water plants have long residence times, highly variable influent rates plus their compositions, and complex kinetics as well as hydrodynamics. The data gaps and flowsheet model shortcomings were consolidated using computational fluid dynamics while influent data analysis utilized neural nets. The methodology is applied to development of operating policies of a Hillsborough County waste water treatment plant.

512936 Comparative Process Analysis Techniques Applied to Four Propylene Oxide Manufacturing Routes

Abstract:

Comparative Process Analysis Techniques Applied to Four Propylene Oxide Manufacturing Routes

Dr. John C. Prindle, Jr., P.E.

Torchwood Enterprises, LLC

Several chemistries are being practiced commercially to manufacture propylene oxide. These include the conventional routes via t-butyl hydroperoxide, ethylbenzene hydroperoxide (Shell's SMPO process), hydrogen peroxide (BASF/DOW), and the proposed direct propylene epoxidation using hydrogen/oxygen mixtures. The challenges of staying competitive in this market require an understanding of the advantages/disadvantages for both your and your competitors' processes.

Comparative analysis of the commercial-scale processes using an available simulation package is one way of gaining this understanding. This task, however, contains major obstacles. Reaction kinetic models have been reported for many of the reactions practiced and proposed. However, many are heterogeneous multi-site models and cannot be directly entered into any of the simulation packages. Even though a statistical range is typically reported for each kinetic parameter in these models, how will the acceptable variation in these parameters impact downstream separations? Are there better separation alternatives which might be less sensitive to these parameter variations? This presentation will discuss the advantages and challenges of applying ASPEN Plus® to the simulation of a commercial-scale process for each of the four routes. Several findings from this comparative analysis will be highlighted.

512725 Application and Challenges of Numerical Simulations in Modeling and Scale-up of Circulating Fluidized Bed Reactors

Abstract:

Circulating fluidized bed (CFB) reactors have a long history of commercial applications ranging from coal gasification and Fluid catalytic cracking (FCC), to newer applications such as catalytic conversion of methanol to olefin (MTO) and even CO_2 Capture using solid sorbents.

Like any other technology, there are challenges in the way of scale-up, deployment and optimization of this technology, based on the lab or bench scale studies. On the other hand, despite of the long history of circulating fluidized bed operations in mature technologies such as Fluid Catalytic Cracking (FCC), complete understanding of the origin and nature of the inherently complex flow structures observed in these devices is still lacking. Adding chemical reactions to the picture makes it even more complex. Therefore, beside experimental studies, advanced and rigorous numerical modeling, i.e. computational fluid dynamics (CFD), is essential to shed light on the complex behavior and flow structure in these systems and to bridge the gap between lab/bench scale and large commercial scales.

In this talk, the strengths and shortcomings of the existing mono-disperse continuum models in simulations of fluidized bed reactors will be reviewed. The needed improvements in the inter-phase exchange properties, frictional viscosity and considering the effect of polydispersity through a novel mathematical approach, will also be discussed.

513072 Intensified-By-Design (IbD): Creating a Platform for Facilitating Process Intensification in Solids Handling Applications

Abstract:

Since its inception, the definition of process intensification (PI) has been broadened from apparatus-centric â€˜miniaturisationâ€™ to the development of processing techniques that offer a drastic reduction in spatial- and energetic-requirements as well as in waste production. With the resulting enhancements in mixing, heat/mass transfer and all other benefits due to PI, many processes involving gas and liquid systems are becoming more proficient in performance and efficiency. On the other hand, the lack of focus in PI for processes involving solids handling has been attributed to the perception of fouling or blockage potentials in micro channels which have often been the technology of choice for miniaturisation. To address this, a clear understanding of industrially-relevant solids handling processes must be undertaken for proper equipment design, thereby remediating the aforementioned hindrances and hence, help facilitate the implementation of PI. These processes have been classified as precipitation/crystallisation, separation, granulation, mixing, particle coating, milling/grinding, catalytic reactions, particle classification, drying, thermal processing and bioprocessing.

In this presentation, five relevant powders- and solids-handling PI technologies are highlighted: (1) The Spinning Disk Reactor (SDR) applies a high centrifugal acceleration to liquids flowing on its surface, forming an extremely thin wavy film with a high surface area to volume ratio. This results in a rapid micro-mixing, creating high supersaturation levels, suitable for precipitation. (2) The Oscillatory Baffled Reactor (OBR) involves a fluidic oscillation in tubes equipped with baffles, resulting in enhanced mixing and heat/mass transfer. This technology has been applied for continuous crystallisation, particle suspension and catalytic reactions. (3) The Taylor-Couette reactors (TCRs) are composed of two differentially rotating concentric cylinders which are separated by a small gap. Flow within the annular space of the Taylor-Couette device experiences a sequence of hydrodynamic instabilities, from laminar Taylor vortex to turbulent regimes, as the rotational speed of the cylinder increased. This ï¬‚ow dynamics is usually characterised by the rotational Reynolds number or Taylor number. The vortex structure in various regimes of hydrodynamic instabilities could be exploited for various potential applications, including particles classification, crystallisation and granulation. (4) Microfluidics deal with the behaviour, precise control and manipulation of fluids that are geometrically constrained to a small scale, typically sub-millimetre. Micro-fluidized beds, for example, utilise the excellent mass and heat transfer characteristics in many applications e.g. kinetics measurement in solid catalysed reactions, detection of infectious bacteria, etc. (5) Twin screw granulator (TSG) is set to transform pharmaceutical processing by offering continuous granulation of drug formulations (currently carried out in batch), thus complying with Quality-by-Design concept, which highlights an automated real-time quality monitoring and control in tablet production.

For these and other solids-related PI technologies investigated in the Intensified-by-Design (IbD) project (funded by the EU Horizon 2020), detailed processing, control and fouling data are compiled in a cloud-based platform. The platform includes built-in design modules, which output datasets for users to facilitate reactor design and assembly, as well as a holistic process optimisation i.e. upstream and downstream intensified unit operations, solids handling capability, cleaning methods and the expected economic and environmental quantitative impacts.

519833 Rapid, In-line Characterization to Support Process Intensification Concepts

Abstract:

Products consisting of processed particulate solids are utilized at some point in the majority of many commodity industries, including the food, pharmaceutical, agricultural and chemical industries. Careful monitoring and control of the manufacturing process is crucial to maintaining high product quality, especially in situations in which process intensification is employed because individual operations often require more stringent specifications and also each operation has greater impact on product properties (i.e. intensification of changes to the product increases the feedback/controls challenge). Particulate solid materials are often much more difficult to handle and feed than liquids or gases, which also exacerbates materials processing difficulties. This talk presents methods to assess material handling and feeding properties of particulate solids, including particle size/shape distributions, density, cohesion, and internal angle of friction (all of which are metrics of flowability). 3D laser scanners and a 2D digital camera are used to automatically measure particle size/shape distributions on a moving conveyor belt, and a custom hopper with adaptable geometry is used to measured density, cohesion, and internal angle of friction. Flow properties are determined by continually adjusting the size of the outlet in a plane-flow hopper to maintain a target fill level in the hopper. The angle of the hopper walls can also be adjusted to tune the geometry to different materials. Test results, including wall pressure measurements, are compared to simulations and indicate that these methods show promise for in-line quantitative estimation of the flow properties of compressible and anisotropic materials.

512700 Hybrid Heat Pipe Screw Dryer: A Novel, Continuous and Highly Energy-Efficient Drying Technology

Abstract:

This work investigated a novel type of screw conveyor dryer (SCD) equipped with a passive heating provided by an annular heat pipe, termed "Heat Pipe Screw Dryer" (HPSD). The device was designed as a compact and continuous drying equipment with an energy efficient operation as its core principle. The heat pipe was firstly characterised, resulting in the lowest thermal resistance of 5.6 x 10^{-3} °C/W at certain conditions: 200°C setting temperature, 11% filling ratio and in (laid flat) horizontal orientation. These conditions therefore, were employed in the drying of a ceramic raw material slurry (initial moisture content: 33 wt. %, wet basis). Moisture reduction of the material increased with increasing temperature and residence time. Approximately 46% moisture reduction (corresponding to 18 wt. % final moisture content) was achieved at 200°C and 260s conditions with the resulting energy efficiency of 2.04 kg/kWh, which places the HPSD as more energy efficient than conventional dryers e.g. fluidised beds, rotary and spray dryers. In addition, thermal analyses showed that the HPSD performed better than a current steam jacketed-SCD, by ~35% higher in the overall heat transfer coefficients. These were attributed to the excellence of the annular heat pipe as a thermal conductor as well as the negation of any pumping requirement of heating fluid. On the other hand, solids buildup (fouling) also increased with moisture reduction, from 3 to 62% for 10 to 46% moisture reductions, respectively. Applying a non-stick "Xylan" coating on the screw conveyor resulted in a ~10% fouling remediation. From experimental data, it was concluded that the fouling could be improved significantly if the entire inner barrel was coated. Reducing the barrel clearance and periodic wash-ups were among other fouling remediation strategies. The HPSD as a multifunctional technology, combining both drying and heating, is a worthy technology to be further investigated, developed and applied, creating a more energy-efficient drying processing.

513193 Process Intensification and Cost Reduction for in-Plant Biomass Preprocessing Equipment

Abstract:

Process intensification and cost reduction for in-plant biomass preprocessing equipment

*Michael Blaise**, *Nicholas, R. Schwartz, Paul E. Yelvington* Mainstream Engineering Corporation, Rockledge, FL *Corresponding Author: mblaise@mainstream-engr.com Abstract

With growing concern about the sustainability and environmental impact of fossil fuels, interest in fuels derived from biomass feedstocks (e.g., agricultural residues, forestry residues, energy crops, and organics in mixed solids wastes) has risen steadily. Designing low-cost feedstock preprocessing systems that can handle a variety of materials is crucial to ensuring efficient and economic plant operation. Conventional preprocessing and solids handling consists of size reduction (i.e., comminution), drying, conveyance, and an inerting system to purge and pressurize the feedstock prior to introduction into a substantially oxygen-free conversion reactor such as a fluidized-bed fast-pyrolysis reactor. In this work, process intensification of preprocessing equipment for thermochemical biomass conversion was designed to take advantage of existing hot flue gas and eliminate the need for separate energy-intensive drying or purging steps. This approach combines drying, knife milling with in-line sharpening, conveyance, and inerting functions. This process-intensified approach is referred to as the integrated DCIMS (drying, conveyance, inerting, milling-sharpening) in-plant feedstock preprocessing. The DCIMS approach eliminated the need for a rotary drum dryer and second conveyor compared to the baseline conventional approach. A better than 5% additional reduction of moisture content was achieved when heating the knife mill with hot flue gas compared to an unheated baseline. Synergy between drying and milling led to reduced heat duty and reduced mill shaft power through combining these processes. The DCIMS approach demonstrated a more uniform (and smaller) milled product without increasing the energy required for milling. Experiments showed that the final moisture content for pine feedstock was 9% (a 15% reduction from a 24% starting moisture), for sugar cane bagasse was 8% (a 20% reduction from a 28% starting moisture), and for mixed waste was 5% (a 25% reduction from a 30% starting moisture). Internal integrated sharpening was also demonstrated successfully. The quality of the blade improved significantlyâ €"feedstock product was more uniform, and the power required for milling decreased by 17%. The estimated operating cost savings from the internal sharpening system was 4%. Heated pneumatic conveyance was also shown to effectively perform biomass drying. A custom motive nozzle for an eductor allowed for low-pressure flue gas to be used directly in the pneumatic conveyance of the feedstock. Integrated DCIMS feedstock preprocessing reduced the operating costs by 37% to $27/dry ton at a 1,000 ton/day scale, based on a technoeconomic analysis (TEA) of the process.

Keywords: Fast pyrolysis, process intensification, preprocessing, biomass, bio-oil

513659 Simulation of Intermediate Boilers in Distillation for Piloting and Scale-up

Abstract:

Distillation processes with impurities that are intermediate boilers between your top and bottom products (non-key lights and heavies) require special attention when scaling up. To minimize risks associated with commercialization of new technologies it is common to test in a lab or even pilot the most complicated parts of the process. This discussion focuses on the complexities encountered with distillation processes that have multiple byproducts and impurities because of their importance to the process design. Careful study of process design parameters in simulations can improve the likelihood of success, and attention to detail is key for low level impurities.

Piloting a distillation system is fairly straight forward but when the system has intermediate impurities the design of the pilot can become complex. Understanding the column profiles for each species can help to avoid complications. The goal is to have a pilot plant that can be used to determine optimum conditions for the design of a larger commercial scale operation. When non key intermediates are present they can build up and cause "burping" or "dumping" in laymen terms. This can occur when the design of the column enables the concentration of non-key lights (in the rectification section) or concentration of non-key heavies (in the stripping section of a column). When the system has a small upset, there can be major impacts to product profiles and the column product streams. A robust design of the pilot can help to identify and correct these problems and study the most effective ways to prevent them from occurring at the commercial scale.

Simulation of the column is a critical starting point to understanding the capabilities of the pilot plant design. Paying attention to the intermediate boiling components and their composition profile can help to identify where to make special arrangements for the column design to enable measurements or in some cases retrofit of the column. Custom designed columns can be expensive and knowing where to expect problems can help minimize costly replacements and or missing the opportunity to study the potential problems before the commercial design.

Process optimization of distillation columns is important for cost of production, cost of operation, and product quality. Simulating the column is not a substitute for piloting, but it can help to identify areas which need further study during the piloting stage. A sensitivity study of the component profiles can guide the piloting design efforts and the experimental run plan in order to identify and minimize scale-up problems.

513877 Adsorption: The Forgotten Unit Operation

Abstract:

Benzene is a much more common chemical than generally supposed. Thus, the opportunity to contaminate a second chemical inventory with trace quantities of benzene is not insubstantial. Such contamination occurs in tank farms and transfer lines. It can also occur in poorly flushed batch and semi-batch reactors. Also, many chemical processes use benzene-containing ligands to activate a catalytic metal. Unfortunately, such ligands decompose into a variety of by-products, one of which is benzene. Such decomposition contaminates the process as well as the desired product with benzene

Removing benzene by distillation generally produces a large benzene-rich by-product stream and a product stream containing low parts per million benzene. Thus, removing parts per million benzene from a product stream is expensive.

Removing benzene via adsorption produces a product stream at or below one part per million benzene without producing a large benzene-rich by-product stream. Adsorbed benzene is recovered by heating the adsorbent, volatizing the benzene-rich hydrocarbon, then condensing the resulting vapor. The benzene-rich hydrocarbon can be sold for fuel or consumed as fuel at the site.

Unfortunately, due to its molecular structure, benzene does not adsorb onto activated alumina or activated carbon. Benzene must, therefore, be adsorbed onto a size exclusion adsorbent, such as molecular sieve.

This presentation discusses using 10X and 13X molecular sieve to adsorb benzene from 1-hexene. We present the benzene adsorption isotherms for each molecular sieve and discuss their implications.

PROCESS INTENSIFICATION OF MULTIPLE SEPARATIONS IN A PILOT PLANT PROCESS VIA THE IMPLMENTATION OF A NUTSCHE FILTER

Anika Coolbaugh, Mid-Atlantic Technology, Research and Innovation Center (MATRIC), South Charleston, WV

Jeremy Rader, P.E., Mid-Atlantic Technology, Research and Innovation Center (MATRIC), South Charleston, WV

Biography

Anika Coolbaugh (presenter) is a chemical engineer at Mid-Atlantic Technology, Research and Innovation Center (P.O. Box 8396, So. Charleston, WV 25303). She earned a BS in chemical engineering and certification in biomedical engineering from West Virginia University in 2017.

Jeremy Rader is the pilot plant supervisor at Mid-Atlantic Technology, Research and Innovation Center (P.O. Box 8396, So. Charleston, WV 25303). He holds a BS in chemical engineering from West Virginia University Institute of Technology and is a licensed P.E. in the state of West Virginia.

Introduction

The intermediate piloting scale between a laboratory process and a commercial plant is an important step that allows for the demonstration and validation of a new process. During scale up, however, it is not uncommon for the process to go through a significant technology change. This can be a result of the infeasibility of the process at the larger scale using the laboratory technology or it can be the result of optimization and process intensification.

The final goal of the scale up process is typically to create a design that would be viable at a commercial scale. While this is certainly a goal in most pilot scale projects, depending on the project specific goals of the customer, there may be other drivers that influence the design. These influences could lead to a design that would accomplish the set goals, but has the possibility of being an infeasible option at the commercial scale.

An example of such a process exists in a pilot plant facility operating in South Charleston, WV. This solids-handling process was originally studied at the lab scale and consisted of a reaction carried out batch-wise in a solvent, followed by a precipitation to separate the product from solution. The final product was then washed and filtered to remove residual reactants before being dried in a vacuum dryer for use in the customer's step 2 process.

Evolution of the Filtration Step

The washing and filtration process went through a number of evolutions in the early stages of process development. The original set up involved the use of a Buchner funnel for a vacuum filtration. The precipitated product was filtered and the resulting cake was removed from the funnel, resuspended in additional anti-solvent to wash, and then filtered again. This wash and filter step would be repeated as many times as was necessary to ensure that all residual reactant was removed. The result of this method was a cake that was still approximately 50 wt% anti-solvent. In addition to the poor separation of the solid product from the anti-solvent, it was also a time consuming process. This was particularly when the solids were of lower quality than average because the low quality product would clog the filter and prevent the anti-solvent removal.

In an attempt to speed up the separation, the filtration was modified and the precipitated product was put under nitrogen pressure in a bag filter to remove the anti-solvent. The product could then be removed and washed in the original method. This method did prove to be a quicker method of separation, however, it was also less effective. The nitrogen pressure failed to push the methanol through the cake because of empty air space at the top of the bag and cracks in the cake that allowed the nitrogen to bypass the solids. This process resulted in a cake with approximately 60-70 wt% anti-solvent. This residual anti-solvent increased the time needed in the vacuum oven for drying, making the overall batch time very similar to the original method.

Both of these original solids handling processes required three separate pieces of equipment. A vessel was needed for the precipitation and a filter was needed for the solid-liquid separation. There also needed to be an easy way to transfer the material between the vessel and filter in order to transfer it for washing and then return it for another separation. Finally, a vacuum dryer was needed to remove the residual anti-solvent.

The original process also required high raw materials usage, was time consuming and manual labor intensive. Despite the excessive anti-solvent usage, the washing step was still not as effective in removing the residual reactants as the customer required for the final product to be used in their step 2 processing.

Project Goals

Although there was still some uncertainty in the final process design, it was decided that the process would be scaled up and constructed at the pilot plant in order to meet the customer goals. During construction, work continued in the lab to provide small scale samples for use in the step 2 laboratory process as well as to optimize the washing and filtering process in order to find a solution that would meet the customer's quality specifications.

There were two primary goals for the scale up of the process in the pilot plant. First, it was necessary to provide scale-up information on the reaction portion of the system in order to validate the chemistry at a larger scale for future commercial implementation. The primary goal, however, was to deliver market development samples as quickly and inexpensively as possible.

Because the focus of the pilot scale was production of product rather than to obtain scale-up data on the solids-handling portion of the system, the opportunity arose to investigate new methods of processing that could reduce the capital and operating costs at the pilot plant scale, even if the final design would not be the ultimate commercial solution.

Simplification of the process was also desired in order to reduce the overall footprint of the process, which was located in a shared space with another process.

Operations of a Nutsche Filter

A nutsche filter is a vessel capable of separating solid-liquid mixtures by various methods. Material is added batch-wise to the filter through feed ports on top of the filter head. The filter is also capable of further processing of the solid by washing and drying.

Spray nozzles in the top of the filter allow for both reslurry and displacement washing. A set of reslurry arms that can be raised and lowered to multiple pre-set positions to aid in the wash step. In the case of a reslurry wash, the arms are lowered to a position just above the filter screen and rotated to provide a gentle agitation to the slurry mixture. For a displacement wash, the arms can be lowered to just above the cake and rotated in the opposite direction tosmooth the solids to aid in even washing over the cake and prevent and cracks or channels that would cause less efficient washing.

The bottom of the filter is a stainless steel screen filter. Nitrogen pressure can be added from the top of the filter to drive the liquid through the filter, below which the liquid is collected and removed from the filter to a waste tank. There is also a flat plate heat exchanger below the filter and a jacket around the filter to provide heat during drying. Vacuum can be pulled on the filter and nitrogen can be sparged up through the heat exchanger to further aid drying.

Older nutsche filters were designed with a discharge auger arm and bottom discharge chute. When the solid product has been separated and treated, the discharge arm is lowered to the bottom of the filter and rotated. As the augers inside the discharge arm rotate, they are able to simultaneously discharge and grind the solid product. The disadvantage to this design is that only about half of the volume of the nutsche filter is usable because filling any more would cause liquid to flow through the discharge arm and into the chute, contaminating the final product. Modern filters are equipped with a side discharge chute, which allows the full volume of the filter to be used, but loses the ability to grind solid product during discharge.

Factors Contributing to the Decision to Implement a Nutsche Filter-Dryer

Many factors made a nutsche filter an ideal technology candidate for this back end process. The versatility of the equipment allowed the entire second half of the process to be contained within a single piece of equipment. This included the precipitation, filtering, washing, drying, and solids handling unit operations.

The filter could be prefilled with anti-solvent to facilitate the precipitation step of the process as the reactor contents were transferred into the filter, eliminating the need for a separate precipitation vessel. Because the solid slurry could be formed inside the filter, it also eliminated the need to transfer any solids or slurries to other unit operations.

The availability of nitrogen from the top of the filter enabled the pressure to be applied evenly over the solids cake in order to drive out the liquid more effectively than what was possible in the lab process.

The possibility to agitate during reslurry washing or smooth the top of the cake during displacement washing allowed for some flexibility in the process as more information was gathered from the laboratory. As the washing step of the process was tested in the lab, some of the minor changes could be implemented in the pilot plant scale in order to improve the final quality of the product.

The ability to dry the product with heat and vacuum from inside the filter and then grind and discharge it using the auger eliminated any need to transfer solids between pieces of equipment. With the difficulties that typically come along with solids handling processes, this was a significant advantage to using the nutsche filter in the process.

In addition to the benefits of keeping all of the unit operations in one piece of equipment, there was also the added benefit of reducing the manual labor associated with the project. The original design for the pilot plant required manual transfer from a centrifuge to a dryer multiple times in order to process the entire batch. The process was also much more manual and was less reliant on the available controls system. Implementing the nutsche filter eliminated much of this manual labor and made the entire process more reliant on process controls, keeping the operators' direct interaction with the process at a minimum to ensure safer operations.

Additional Learnings from Lab Scale Testing During Construction of the Pilot Plant

Throughout the construction of the pilot plant process, work continued in the laboratory to optimize the process in order to produce higher quality product. The reaction was modified to create a semi-continuous process as opposed to the original batch process. Although the quality increased, the washing step needed to be the main focus in order to produce product that was of adequate quality for use in future processes.

To replace the bag filter in the process, a washing bag centrifuge was implemented into the process. This modification significantly increased the effectiveness of the washing step, removing almost all of the residual reactants from the product while also reducing the amount of anti-solvent needed by doing a cake wash rather than a reslurry wash. The centrifuge also significantly lowered the amount of residual anti-solvent remaining in the solid product after washing, shortening the drying time of the product.

Although a cake wash was implemented into the pilot plant process, these findings indicated that the nutsche filter was unlikely to be the final commercial solution. The nutsche, while far more efficient than the bag filter that was part of its original design bases, would not be as efficient at washing or removing the residual anti-solvent as the centrifuge. The nutsche filter would, however, still be able to meet the production requirements set by the customer.

Results of Initial Nutsche Filter Testing

The initial validation of the pilot plant technology consisted of two reactor batch tests. Both runs ran well in the reactor and produced the amount of polymer that was expected. The quality of the first batch was similar to what had been seen in the batch runs with the bag filter set up. This meant that there were still high levels of the residual reactant leftover in the batch. The spray nozzles did not work as they were expected to and provided a stream of anti-solvent rather than a spray. This may have decreased the efficiency of the washing cycle and could be corrected to see a higher quality final product.

The second batch resulted in very poor quality due to poor precipitation of the product. The first batch had over filled the nutsche filter, resulting in liquid pouring through the auger and into the discharge chute. In order to attempt to correct this, the amount of anti-solvent used in the precipitation step was reduced. As a result, the product did not properly precipitate out of solution, leaving a wax-like product in the nutsche filter that conglomerated into large clumps rather than the expected fine powder. The clumps were too hard for the augers to properly discharge, causing clogging in the auger arm and leaving the nutsche filter unable to be automatically discharged.

In order to discharge the material, the manhole on the filter needed to be opened and the material had to be scooped out of the filter by hand into buckets. Removing the material and cleaning the filter so that it was ready for a new batch took just over two days with two operators on the job.

Benefits of Implementing a Nutsche Filter-Dryer

There were multiple benefits to the implementation of the nutsche filter-dryer in this solids handling process. The ability for a single piece of equipment to be utilized in so many different unit operations makes the construction of the process simpler and decreases the overall process footprint. It also decreases the labor requirement for the process by eliminating the need to transfer any solids or slurries between unit operations and by making the process more reliant on automated process control.

Disadvantages of the Nutsche Filter Implementation

Although the nutsche filter would theoretically be an excellent option for this solids handling process, there were some draw backs to its implementation. First, the capital investment for a nutsche filter is fairly high: approximately $200k for a used filter or $450k for a new filter of 1 m^2 area. It is possible that you could implement all of the separate unit operations that a nutsche filter replaces for a lesser or equal investment.

Another drawback to the implementation of the nutsche filter is the maintenance time required when there is a process upset. Maintenance of a nutsche filter is more complex than the maintenance of the individual unit operations would be. This means that more time needs to be spent inactive in order to properly clean and prepare the filter for future runs after an upset.

One of the largest disadvantages to the nutsche filter are that the equipment does not lend itself to a continuous process. Because all of the operations are contained within a single piece of equipment, there is no way to continuously feed to a nutsche filter. In order to implement the filter into a continuous process, it would be necessary to install a holding tank for the reactor contents and, depending on production needs, multiple filters would need to run in parallel.

The other major disadvantage is that the filter is not particularly effective at any one unit operation. While it is capable of precipitating, washing, filtering, and drying, each of these tasks is better achieved by its own individual piece of equipment and separating the tasks would most likely result in a final product of a higher quality than can be achieved in the nutsche filter-dryer.

Conclusion

Based on the information that was available at the time of the design and construction of the pilot plant unit, the nutsche filter performs well in the process and achieves the customer's production goals. It produces product that is of similar quality to the early lab experimentation, meaning the wash step is inefficient and leaves behind an excess of residual reactant.

Although the filter continues to meet the customer's production goals, with new information from the laboratory process, it is clear that this is not the commercial solution. The implementation of individual pieces of equipment for each unit operation would provide a more efficient process that would be applicable to continuous operation and would produce higher quality product.

512932 Overcoming Heat Transfer Bottlenecks in Non-Equilibrium Chemical Processes through Autothermal Operation

Abstract:

Overcoming Heat Transfer Bottlenecks in Non-Equilibrium Chemical Processes through Autothermal Operation

Robert C. Brown[1, 2] and Chad Peterson[2]

[1]Bioeconomy Institute

[2] Department of Mechanical Engineering

Iowa State University

Innovations in Process Intensification Session

Topical C: Process Intensification

AIChE Spring Meeting, Orlando, FL

April 22-26, 2018

Abstract

Heat transfer is the bottleneck in many chemical processes, limiting throughput for a reactor even when chemical reaction and mass transfer are very fast. Efforts to enhance heat transfer can add significantly to equipment costs. Furthermore, with heat transfer rate limiting, reactor capacity only scales as the square of reactor diameter, constraining throughput as reactors are built larger and increasing capital costs faster than expected from economies of scale principles.

Autothermal operation, which balances the energy demand of endothermic chemical reactions with energy supply from exothermic chemical reactions, can overcome heat transfer bottlenecks. Although autothermal operation has been previously demonstrated for *equilibrium* chemical processes (prominent examples include methane steam reformers and biomass/coal gasifiers), application to *non-equilibrium* processes has been little explored. Identifying complementary endothermic and exothermic reactions that can occur simultaneously while preserving the same products of an externally heated (or cooled) chemical process is not trivial. However, the gains in *process intensification* would be substantial, increasing throughput and reducing capital and operating costs of a chemical process.

We have experimentally demonstrated the principle of non-equilibrium autothermal operation for the fast pyrolysis of biomass. Autothermal pyrolysis was achieved by admitting a small amount of air into a fluidized bed pyrolyzer, which partially oxidizes biomass reactant or pyrolysis products to provide the enthalpy of pyrolysis without substantially affecting the desired pyrolysis products. Decreases in non-condensable gases such as carbon monoxide and methane during autothermal pyrolysis would suggest their oxidation is responsible for a significant fraction of the exothermic energy released during this process. However, autoignition temperatures for the non-condensable gases in pyrolysis products is higher than typically occurs in fast pyrolysis, suggesting other exothermic chemical reactions provide the energy to support autothermal pyrolysis.

We are developing a chemical kinetic model of autothermal pyrolysis to better understand the source of energy for the process. The model indicates that non-condensable gases are not substantially oxidized during autothermal pyrolysis. Instead, precursors of these gases along with other light oxygenated compounds are oxidizing and providing the enthalpy for pyrolysis. The model is also investigating solid-gas reactions to investigate the contribution of lignin and char to the energy required for autothermal pyrolysis.

512574 Reverse Flow Reactors: Process Intensification for High Temperature Chemistries

Abstract:

Global chemicals demand is expected to surpass 200 million metric tons by the end of this decade, with a growth pace faster than the global GDP. The precursors to these chemicals are often manufactured using high temperature endothermic reactions such as reforming and pyrolysis. Substantial energy is consumed to heat feed streams to reaction temperatures, and ineffectively captured as product streams are further processed. Additionally, the significant capital, physical and carbon footprint requirements make these processes ideal for a process intensification (PI) play.

We describe a reverse flow reactor (RFR) approach that efficiently integrates heat-transfer, chemical reactions and cyclical in-situ de-coking by alternating flows between reaction (endothermic) and reheat (exothermic) steps in a fixed bed reactor. The use of high surface area, thermally robust ceramic internals allow operating temperatures up to 1500°C enabling higher conversion and short residence times (higher selectivity) long sought out for optimal performance.

This presentation will provide a brief introduction to RFR, discuss the key reactor features and exemplify a few applications. Finally, the need for a multi-disciplinary approach involving interplay of modeling, materials development, and experimental demonstration is advocated for the advancement of new PI technologies.

512776 Thermodynamic Modeling for CO_2 Absorption in Aqueous Potassium Carbonate Solution with Electrolyte NRTL Model

Abstract:

Potassium carbonate has recently gained attention and is considered favorable for CO_2 removal due to numerous reasons such as low economic costs and energy consumption, low volatilization and degradation and less environmental impacts compared to amine solutions. Therefore, we develop a comprehensive thermodynamic model for the $K_2CO_3 - H_2O - CO_2$ system using the Electrolyte Nonrandom Two-liquid (eNRTL) activity coefficient model in conjunction with PC-SAFT equation of state. Vapor-liquid equilibrium (VLE), heat capacity, excess enthalpy, mean ionic activity coefficient and osmotic coefficient data are used to determine eNRTL temperature dependent binary interaction parameters for both $K_2CO_3 - H_2O$ binary and $K_2CO_3 - H_2O - CO_2$ ternary systems. The model accurately represents all the thermodynamic properties and covers a wide temperature range from 273.15 to 473.15 K and CO_2 loading from 0 to 1 The model can be merged with other amine models for mixed K_2CO_3 – amine solutions and the model can be used for comparison with other CO_2 capture chemical solutions. The model should be a useful tool in the process modeling, simulation and design of CO_2 capture processes involving hot, concentrated K_2CO_3 solutions at elevated temperatures and pressures.

512813 Single and Multi-Objective Optimizations Using Parallelized Process Simulators

Abstract:

Equation-oriented (EO) optimization methods are difficult to apply to rigorous thermodynamic methods, discontinuous variables, or systems where the number of equations is not constant. Despite these challenges, EO methods are still quite commonly used for process optimization, due in part to a perception that EO methods are faster than process simulation-based methods.

By integrating parallel processing techniques into modern process simulation and optimization, rapid results can be achieved comparable to the speed of more traditional EO modeling methods. These speeds can be attained even while using first-principle models, without risking validity of the model through too many simplifying assumptions. In this presentation, several process examples will be considered to explore the advantages of using simulators for optimization projects.

512830 Thermodynamic Modeling of the $LiCl+LiBr+LiNO_3+LiI+H_2O$ System with the Electrolyte NRTL Model

Abstract:

Thermodynamic models are essential to facilitate the advancement of process design, optimization, and operation range of absorption heat pumps and refrigeration units that utilize lithium salt systems. The lithium bromide-water solution is a popular working fluid due to its environmentally friendly, nonvolatile, and non-toxic properties; however, this solution has issues with corrosion and crystallization. One proposed solution is to modify the lithium bromide-water working solution by adding anticrystallization and anticorrosion agents such as lithium chloride, lithium iodide, and lithium nitrate. The objective of the present work is to develop a comprehensive thermodynamic representation of the $LiCl+LiBr+LiNO_3+LiI+H_2O$ quinary solution by implementing the symmetric electrolyte NRTL (eNRTL) activity coefficient model. This solubility model correlates the nonideality of the solution with the composition dependency by utilizing two binary interaction parameters for each of the electrolyte-electrolyte and molecule-electrolyte pairs in the system. Each parameter is determined by applying up to three temperature dependent coefficients and then regressed from experimental data. This allows the parameters to establish the accuracy of the model for concentrations up to saturation and temperatures between 273.15-473.15 K.

517124 Building with Blocks

Abstract:

Chemical processing facilities traditionally use a â€œstick buildâ€ approach for expansions, retrofits and new plants. This involves field installation and erection of all equipment, piping and components. Costs for this strategy are well understood which perpetuates its continued preference. However, this may not always be the most cost effective approach or the most appropriate strategy. In some instances Modular Manufacturing offers a lower cost alternative to conventional methods. Modules can be thought of as the building blocks from which processes can be built. In small scale manufacturing these modules can be designed with considerable flexibility and bolted on to for enhanced capability. In some cases modules can be assembled as a faster and lower cost way to build a large facility, particularly when local skilled resources are a limitation.

This presentation will highlight beneficial applications of modular manufacturing utilized in a centralized facility. The presentation will also include identified knowledge gaps that limit a broader use of the technology. More cost information and understanding is needed to know when the best approach is to build with blocks or sticks.

517126 Critical Factors and Criteria for the Development and Deployment of Modular Chemical Processes

Abstract:

The allure of implementing modular chemical processes is the possibility of achieving significant cost savings compared with traditional individually-engineered and stick-built processes. In theory a factory-built, modular process could have a lower construction cost than a stick-built process; however, factory-built processes must be constructed of units that are a size that can economically be shipped and connected in the field. For processes above a certain size, this would necessitate the traditional engineering scaling factors being broken via advances in technology, manufacturing, and engineering design approaches. While the concept of producing modular processes is appealing, overcoming these traditional scaling factors is challenging because current supply chain and manufacturing systems are geared towards the stick-built approach for chemical process equipment. These cost savings will only be achievable if multiple standard units of the same design are produced. System standardization and minimization of one-time engineering and maintenance costs are essential to minimizing total cost of ownership of a modular system.

Considering these constraints and factors, purification processes for wastes or feed preparation are low-hanging fruit for the implementation of modular chemical processes for both distributed and centralized applications. Implementation of modular processes for distributed and centralized locations have different primary challenges. For the case of centralized applications ensuring ease of integration with the existing infrastructure, balance of plant, and controls becomes paramount, while for distributed applications it is essential that the process is highly autonomous, robust, and self-sufficient.

This talk will summarize and examine the critical factors and criteria for the development and deployment of modular chemical processes. Topics to be discussed will include when to choose modular, system size selection, process controls, standardization, manufacturing, and design for CAPEX and OPEX minimization. A focus will be placed on examining these factors relative to IntraMicron's recently developed Sour Gas Sweetening Process, SourCat^{TM}. The SourCat^{TM} desulfurization process is currently being commercialized by SourGas LLC, a wholly owned subsidiary of IntraMicron. The SourCat^{TM} process will be deployed commercially as distributed modular units in the Oil and Gas Industry. The transition of this process from a modular pilot plant to a modular chemical process is currently the subject of the RAPID program "Modular Catalytic Desulfurization Units for Sour Gas Sweetening."

517136 Continuous Processing of Specialty Chemicals â€“ Opportunities with MCPI

Abstract:

Specialty chemicals and small commodities market is challenged with relatively small throughputs and relatively high production cost. Dissemination of the costs is prerequisite for making business decision for capital investment in new capacity. The new plant must be near to the market and be able to adjust rapidly to market changes. It must also have possibility for flexible production e.g. change of the product must be relatively simple. Fulfillment of these demands is challenging for the process engineering. In order to mitigate these challenges, design of cheaper, intensified modules, are of most importance since the solution may lay in modular plant and/or in process intensified equipment such as flow reactors or dividing wall columns. In order to discover and utilize the opportunities a dissemination of capital expenditure is essential.

513873 Development of a Two-Stage Thermal Deoxygenation Process for Fuels and Chemicals from Biomass

Abstract:

Thermal DeOxygenation (TDO) is the thermal decomposition of mixed salts of levulinic and formic acids that can produce hydrocarbon oils at yields approaching 80% of theoretical. These acids can be sustainably produced in stoichiometric quantities by acid hydrolysis and dehydration (AHDH) of cellulose. The TDO reaction was scaled up to a 50 L semi-batch reactor to produce more than 20 L of oil for fractionation, blending and hydrotreating. The products were thus demonstrated as suitable blendstocks for gasoline, diesel, jet fuel, and fuel oils. While scale-up facilitated kilogram quantities of products for testing, the carbon loss to char increased from 8.5% at the bench scale to 48% at 50 L floor scale. Isothermal reaction experiments have identified two reaction regimes: 1) a low temperature addition-condensation regime that is diffusion-controlled and 2) a high temperature ketonic decarboxylation regime that is reaction-controlled. Through this understanding, a two stage reaction process has achieved yields in the floor scale reactor similar to early bench scale results. Current process development efforts involve testing equipment to move from semi-batch to continuous operations and scaling up the TDO process to match the output of UMaineâ€™s one-ton-per-day AHDH pilot plant. The combined AHDH-TDO process has the potential to produce a suite of bio-based chemicals in addition to fuels, so additional results include techno-economic analyses which consider the potential for co-product revenue enhancement of renewable fuels.

513289 Development of Green Pathways for the Extraction of Medically Active Components Form Grape Waste

Abstract:

Medical sector growth continues to advance thanks in part to a boom in the biomedical industry. Increased technological advancement of this level has afforded people health and wellness not seen in previous generations leading to longer more productive lives. Growth in the Biomedical can be amenably paired with green technologies to further the field in an environmentally friendly way. Innovation in the use of natural components in health fields as a way of improving health is a most promising direction but is one existing with many bottlenecks. Despite the challenges, these technology shifts are necessary to stay in line with environmental regulation, FDA regulation, and public opinion.

The use of Supercritical CO_2 is growing more and more common in many fields. It can be used singularly, with co-solvent, or as part of more complex systems to process Biomass in a more environmentally friendly way. Waste biomass can be especially useful here. What was once not worth processing and was relegated trash or burning can be processed to obtain high valuable compounds. Grape waste from wineries and juicing facilities consists of seeds, stems, rings, meat, and leaves. Most of this waste has little to no value and is relegated to trash or animal feed. With some creating processing, this waste can be a source of valuable oil as well as high-value antioxidants, polyphenols, and flavanoids. Scale is important in these operations and thus a pilot study is being conducted on this process to find the range of capacities that are profitable for this endeavor.

513562 Efficient Numerical Optimization of an Aerated Mixing Vessel for Fermentation

Abstract:

Oxygen is a critical component of aerobic fermentations and is sparingly soluble in aqueous solutions. Due to the slow mass transfer of oxygen from the bubbles into the liquid the rate of cell metabolism depends on the rate of oxygen supply from the gas phase. Mass transfer across gas-liquid interfaces plays also a vital role in many other reaction systems. The mass transfer is a function of the gas-liquid surface area which is a function of the coalescence and breakup rate which is again a function of the local velocity and turbulence. The local flow condition are finally determined by the reactor geometry, the fluid properties and â€" in this study â€" by the rotational speed of the impeller. So from a reactor performance standpoint, high aeration rates and high agitation rates are favorable.

On the other hand the operating costs are mainly determined by the total power consumed for the aeration and by the impeller of an aerated mixing vessel and should be as low as possible.

Traditionally dimensionless correlations and expensive experimental investigations are used to design stirred tank fermenters in production scale. However, design methodologies available for stirred tank fermenters are often insufficient and are using simplifications which affects e.g. the mass transfer significantly. Using computational fluid dynamics (CFD) based on validated models enables using fewer assumptions on the geometry and physics resulting in a better insight into the system.

In this work the performance of a stirred tank fermenter is going to be optimized using CFD simulations in combination with design explorations. After validating some implemented models especially to incorporate bubble size distribution including coalescence and breakup in the finite volume code STAR-CCM+, HEEDS is used to explore the design space. HEEDS is a multi-objective optimization tool using hybrid, adaptive search strategies to find better designs, faster and couples seamlessly to STAR-CCM+, so that no model simplifications are needed. Details how this approach works and how it can be applied to other problems will be shown. Finally the obtained Pareto front w.r.t. power consumption and mass transfer rate will be presented.

513225 Multi-Oil Product Pipeline Scheduling with Appropriate Handling of Oil Mixing Interface

Abstract:

Pipeline is more efficient and reliable to transfer large amounts of petroleum products than other ways via road, railroad, and vessel. To save capital and operating costs, one pipeline needs to transfer different types of oil products to different oil depots according to their ordinary demands. As two different consecutive oil products inside a pipeline will not be separated by a physical divider, the product mixing is inevitable near the interface of two oil products. To guarantee the oil quality adopted by storage tanks at depots, a considerable amount of mixing oil product from the pipeline needs to be dealt with properly at depots. In practice, three kinds of methods are used. The first one is to blend mixture oils into pure product, which might have some oil quality problems; the second one is to separate the oil mixture into pure oil products through distillation; and the third one is the hybrid method coupling the former two methods together. Conceivably, the multiproduct pipeline scheduling problem with consideration of dealing with oil mixing inside the pipeline present a very challenging problem to industries. There are few works reported in this area (e.g., Rejowski et al. (2003); Cafaro and Cerdá (2004 and 2010)). However, they still lack studies on dealing with the oil mixture product.

In this work, a continuous-time based MILP model is developed to optimize the transportation and distribution of multi oil products, which includes a long-distance pipeline, several depots with storage tanks and mixing tanks dedicated to hold pure product and mixture respectively. The oil mixing interface inside the pipeline is modeled as three parts: two product-rich mixing parts on both sides of the mixing interface and one well mixing part in the middle. The hybrid method for blending product-rich mixture into the pure product and refining the well mixing mixture has been adopted. Overall, the scheduling model could provide optimal solutions including product feeding into pipeline, product movements inside pipeline, discharging product to multiple depot tanks, as well as sending out oil product based on consumer demands in the continuous time formulation base. The efficacy of the developed scheduling model is demonstrated by multiple case studies.

Keywords: Multiproduct Scheduling, Pipeline, Mixing Product Management, MILP

References:

Cafaro, D., & Cerdá, J. (2004). Optimal scheduling of multiproduct pipeline systems using a non-discrete MILP formulation. Computers and Chemical Engineer- ing, 28, 2053–2068. ISSN 0098-1354.

Cafaro, D., & Cerdá, J. (2010). Operational scheduling of refined products pipeline networks with simultaneous batch injections. Computers and Chemical Engineer- ing, 34, 1687–1704 .

Rejowski, R., & Pinto, J. (2003). Scheduling of a multiproduct pipeline system. Com- puters and Chemical Engineering, 27, 1229–1268 .

513699 Advanced Model-Based Dynamics Optimization of a Commercial Natural Gas Recovery Unit

Abstract:

Natural gas liquids (NGLs) are the main contributor to the profitability of natural gas processing plants. During the recent period of relatively low gas prices NGLs spot prices have experienced a slower drop remaining relatively steady. Therefore, driving operating plants towards higher NGL production can lead to a more profitable position. In this work, we look to use model-based optimization on a selection of NGL plants to optimize them towards higher profitability and/or lowest energy consumption.

Such studies are meaningful only when based on sound, high-fidelity models of the plant, ensuring that â€œoptimalâ€ solutions obtained satisfy all important safety and operability constraints. In the typical workflow a model is built in commonly used flowsheeting tools based on the Sequential Modular approach. Limitations in this approach result in engineers having to resort to manual trial-and-error simulations with no guarantee of obtaining an optimal solution or even one that satisfies all the important constraints. In this work models are being built in an Equation Oriented environment which allows the use of formal mathematical optimization methods to guarantee an optimal solution satisfying all constraints is found.

In this paper we use the example of an on-going study of a NGL Plant in Abu Dhabi region to explain the practical benefits of Advanced Model-based Optimization. We will describe and show how the key first step in such a study is validation: ensuring that model predictions of key results are matching the actual plant behavior. We will describe how the model is then used to understand the best operational strategies for the facility through model-based optimization. This demonstrates how the approach used overcomes the limitations of the typical workflow and models used. In particular, the use of tools based on the Sequential Modular approach is limited in optimization capabilities due to its problematic handling of non-standard specifications, closing of recycle loops and the unavailability of gradients. The trial-and-error simulations used in this approach are not only time consuming and error prone but often limited in the operating decisions and constraints that can be considered. We will show how using an Equation Oriented environment overcomes these limitations and allows an optimal solution satisfying all operational and safety constraints to be found using mathematical optimization methods.

Finally we will show how this leads to the set-up of operating guidelines to allow operators to drive the plant towards optimal operation; giving them the ability to respond to changes in the plant or economic conditions quicker and with more confidence than previously possible. In addition to moving the plant towards optimal operation, the models will also provide the ability to address a range of operational situations.

513652 CFD Simulation of Gas-Liquid Flow in Monolith Channels

Abstract:

Multiphase monolith reactors are a kind of novel process intensification reactors which are promising for many catalytic gas-liquid-solid reactions encountered in chemical, petrochemical, biochemical, and environmental processes. Multiphase hydrodynamics and distribution characteristics in a monolith bed are the important factors which affect gas-liquid mass transfer in monolith channels significantly and therefore the whole reactor performance. In this paper, a two-dimensional (2D) CFD model has been built for analyzing the distribution characteristics of gas-liquid flow within a monolith bed with 0.048m diameter and 400 cpsi. The volume of fluid (VOF) multiphase flow model is implemented to calculate the flow field and capture the interface between the gas and liquid phases then obtain hydrodynamic parameters in each channels of the monolith bed. The simulation results matched well with the experimental results presented by Zhou (Yuan Zhou, et al, 2014, AICHE Journal).

513347 Comparison of Several Indirect Liquefaction Pathways to Fuels and Co-Products Via Biomass Gasification and Synthesis

Abstract:

Many pathways to convert biomass to fuels and chemicals have been proposed over the years. The economic evaluations of these processes are sensitive to the underlying assumptions and cost-basis year used to estimate production costs making it difficult to compare pathways on an equivalent basis. These differences are inevitable as new information is assimilated into process models and as cost assumptions, such as the cost of biomass, change to reflect the current understanding of the supply chain. Over the last several years, the Bioenergy Technology Office (BETO) of the US Department of Energy used a minimum fuel selling price (MFSP) target of $3 per gallon gasoline equivalent as a program objective to be achieved through research. In petroleum refineries, the co-production of higher value chemicals is an important factor for achieving profitable margins for the overall refinery. As with refineries, co-products with values greater than transportation fuels can improve the economics of a biomass-to-fuel process if the trade-offs with increased production costs are acceptable. Recent techno-economic analyses of several processes to convert biomass to fuels and chemicals via gasification, synthesis, oligomerization and hydrogenation were compared using a common set of modeling assumptions for a cost-basis year of 2014. The analyses were done for state-of-technology and goal cases for process improvements that may be achievable through research. Butanol, 1,3-butadiene, isobutene, LPG, and ethylene were evaluated for their potential to decrease the minimum fuel selling price (MFSP) for gasoline and distillate fuels. The impact of co-products on the MFSP is highly dependent on their market value which can fluctuate significantly year over year. The results of the study showed that within the recent ranges of market prices for the chemical co-products, the MFSP target of $3/gge could be achieved in some cases when chemical market prices were in the upper end of the price ranges and with a relatively large proportion of co-products (up to 50% of total product).

513536 Techno-Economic and Environmental Life Cycle Assessment (LCA) of Biodiesel Production Process from Waste Cooking Oil Via Supercritical Transesterification

Abstract:

This work describes the design of a non-catalytic biodiesel (Methyl Oleate) production process from waste cooking oil (Triolein) using supercritical methanol and propane. As compared to a conventional supercritical transesterification route, this process uses propane as a co-solvent to decrease the temperature â€" from 350Â°C to 280Â°C, reduce the pressure â€" from 400 bar to 128 bar and decreases the methanol to oil ratio - from 1:42 to 1:24 [1-3]. Different plant capacities ranging from 8000 ton/year to 125,000 ton/year were simulated in Aspen Plus software to conduct the techno-economic assessment for a plant situated in the Midwest region of the US. On the basis of mass and energy balance, process equipments such as multi-tubular reactor, evaporator, distillation column, etc. were sized and its subsequent economic analysis was performed in Aspen Plus.

Environmental Life cycle assessment of the process (LCA) was conducted using GREET and GaBi software tools to determine Greenhouse gas (GHG) emissions and evaluate the carbon footprint of the biodiesel production process. The technical, economic and environmental assessment results of this work were compared with conventional acid and base catalyzed transesterification processes to determine the economic and environmental sustainability of the supercritical biodiesel production process.

References:

1. Van Kasteren; J.M.N; Nisworo, A.P, A process model to estimate the cost of industrial scale biodiesel production from waste cooking oil by supercritical transesterification, Resources, Conservation & Recycling, 50, 442-48, 2006.
2. Cao W; Han H; Zhang J, Preparation of biodiesel from soybean oil using supercritical methanol and co-solvent, Fuel, 84, 347-51, 2005.
3. Farobie, O.; Matsumura, Y, State of the art of biodiesel production under supercritical conditions, Progress in Energy and Combustion Science, 63, 173-203, 2017.

513565 Continuous Cultivation of MicroalgaÂ *Nannochloropsis Oculata* to Extract and Characterize Polar Lipids for Cosmetic Applications

Abstract:

Global demand for natural and renewable products have skyrocketed recently, leading to many new possibilities and markets. These new markets have shifted the trend towards the production of sustainable products. Algae, specifically microalgae, that are promising sources of natural ingredients owing to their cell mass containing polysaccharides, proteins, lipids, pigments, and vitamins have been the benign raw materials. Particularly the interest on phospholipids, a subgroup of polar lipids, has been on the rise due to the fact that they are the appropriate candidates to the pharmaceutical and cosmetics industry for various applications such as emulsifiers, liposome formers, solubilizers, wetting agents, and bioactive compounds. Liposomes made out of pure phospholipids are the major contributers to the cosmeceuticals, which can contain up to 20% (w/v) phospholipids, being a part of a $62 billion/year industry in US alone. Besides, cosmeceuticals account for over 12% of the cosmetics market.

Phospholipids are currently extracted from food sources; however, this practice raises sustainability issues because of the competition with food. Microalgae can serve as a source of phospholipids on a more sustainable basis. In order to advance the status of microalgae bioproducts for cosmetics applications, the marine microalga *Nannochloropsis oculata* (*N. oculata*) was cultivated in a vertical flat panel photobioreactor (VFPPBR) in batch and continuous mode. Batch cultivation of *N. oculata* was performed to optimize the growth parameters, estimate its maximum specific growth rate (μ_{max}), and determine its phospholipid productivity. Because the total lipid productivity is essential to increase the phospholipid productivity, optimization of volumetric microalgal biomass productivity in turn boosted up the overall lipid productivity. The measured μ_{max} was utilized in selecting the dilution rate for designing and operating a VFPPBR-based continuous cultivation of *N. oculata* that was deemed to be more scalable and productive operation from a commercialization aspect. Cell harvesting *via* centrifugation was followed by solvent/pressurized solvent extraction to capture and isolate the microalgal polar lipids, specifically phospholipids. The experimental findings from continuous cultivation of *N. oculata* species, the robustness of the strain, and its polar lipid productivity, photobioreactor design, performance of the species in batch and continuous mode of operation, and green pathways for downstream processing to isolate, concentrate and fractionate microalgal phospholipids were analyzed and discussed.

513583 Synergistic Utilization of Coal Ash and Biomass Ash Rice Husk Ash for Zeolite and Sorbent Production

Abstract:

A novel continuous process for zeolites production from coal fly ash and biomass ash (including rice husk ash and oat hulls ash) is reported in this paper. The advantages of combining coal ash and biomass ash include 1) flexible Si/Al ratios adjustment; 2) trace elements such as iron and rare earth metals in coal ash can enhance the catalytic effects for certain zeolites. A comparison of batch process and continuous process is presented.

515795 Conversion of Syngas into Liquid Hydrocarbons

Abstract:

TDA Research Inc. is developing a continuous, high-pressure homogeneous catalytic process to convert syngas derived from biomass gasification into valuable hydrocarbons including alcohols, acetates, and acids with high selectivity and minimum methane formation. Our process uses metal carbonyl catalyst precursors that form catalytically active anionic carbonyl and metal-carbonyl hydrides. The carbonyl anions catalyze CO-transfer reactions, hydride transfer, and hydrogen reduction, and the carbonyl hydrides transfer both CO and hydrogen. They also exhibit water gas shift activity, which allows ethanol to be made using syngas from biomass gasifiers that have hydrogen to carbon monoxide ratios of one or lower. The rate of ethanol formation exceeds 0.26 g ethanol/g metal catalyst/h, which on a metals basis is higher than the best heterogeneous catalysts reported in the literature (K promoted MoS_2), but with the homogeneous system shows much lower selectivity to methane, a low value product. We also present data for proof of concept for a process that includes methanol and catalyst recycle.

Besides forming methanol and ethanol without the need for added H_2, our catalyst promotes the formation of acids and acetates through the esterification reaction as well as catalyzing numerous reduction and hydroformylation reactions. With water-gas shift, $H_2O + CO = H_2 + CO_2$, we have demonstrated that no gaseous H_2 need be added to the reactor to form, for example, methanol and ethanol. Recent results have shown the ability to selectivity promote esterification (producing acetates) in addition to carbonylation followed by hydrogenation (producing alcohols). In addition, our homogeneous catalystâ€™s water-gas shift activity has the potential to minimize residual water in the products leading more energy-efficient product separation.

513795 Hydrogen Production from Biomass

Abstract:

According to International Energy Outlook (2016), global energy demand is expected to increase by 48% by 2040 due to the rapid growth of population, urbanization and economic. The major energy supply is currently reliant on conventional energy resources such as coal (21%), petroleum (28%) and natural gas (32%), which have negative environmental impact i.e. greenhouse gas (CO_2) emissions, air pollution (SO_x, NO_x, particulates and toxic metals etc.). Biomass gasification has attracted attention due to the versatility of its products (hydrogen or syngas: a mixture of hydrogen and CO) that can be used for energy, fuel and chemicals. Hydrogen based economy has been of great interest worldwide. In this study, a two stage gasification of waste wood was carried out to optimise conditions for hydrogen production. The effects of particles sizes and operating conditions such as temperature and steam to carbon ratio etc. on the properties of gas products were investigated. It was found that the gasification temperature and steam to carbon ratio had a significant effect on the properties of syngas. Up to 54.5%mol of hydrogen can be produced in a temperature range of 900-1000°C, providing a ratio of H_2/CO of 1.4. Increasing the ratio of steam to carbon increased hydrogen production up to 64.5%mol and ratio of H_2/CO of 2.94 with small concentration of CH_4 in the gas stream (<2%mol)â€

513857 Model-Based Development of High-Purity Hydrogen Production Unit for Hydrogen Refueling Stations

Abstract:

A conceptual high-purity hydrogen production unit (100 kg hydrogen per day), comprising methane-steam reforming, water-gas shift reactors and pressure-swing adsorption process, for hydrogen refueling stations was developed by the model-based methodology. The system should satisfy the main prerequisites for actual application such as 1) higher than 70 % hydrogen production efficiency on the basis of higher heating values and 2) lower than 10 ppm of CO concentration at the outlet of the CO removal process. The heat exchanging system between heat sources and sinks in the reforming process was optimized by the commercial process simulation software package and the heat and mass transport phenomena in the steam-methane were also intensively studied by the computational fluid dynamic models. In addition, the optimal vacuum pressure swing adsorption process was also developed to purify the hydrogen from the product gas of the water-gas shift reactor. Thus, the developed system was tested and the operating data was compared with the mathematically predicted values so that the fidelity of the model-based process development techniques is evaluated. The proposed model-based process development techniques will contribute to the optimization and scale-up of the process.

Designing for Innovation: Process and Technology Configurations for Oil Sands Production

Evar C. Umeozor, Ian D. Gates
Department of Chemical and Petroleum Engineering, University of Calgary

Abstract

Bitumen from the oil sands is an unconventional oil resource and the major source of Canadian oil production, representing about 70% of total daily volumes. However, the high viscosity of the oil and the nature of its deposition necessitate the use of special tools and recovery techniques that are currently energy, water and emissions intensive. Recent lows in the price of crude oil and evolving environmental conservation goals call for the development of processes and technologies to improve operational performance. This work presents a mixed-integer linear optimization model aimed to simultaneously improve three operational performance indicators representing the economic, emission, and energy intensities of the processes and technologies deployable in a Greenfield facility. The proposed model identifies Pareto-optimal facility configurations for in-situ heavy oil recovery from the oil sands. The optimization criterion is formulated as a weighted sum of the performance indicators of the facility. The performance attributes are captured as the greenhouse gas emissions, process heat and electricity requirements, and the capitalized costs of the facility. Facility operation is constrained to account for field conditions via governing physical equations and other policy-related requirements. By identifying individual unit operations in the bitumen recovery process, covering both the surface and sub-surface elements, existing challenges/limitations of current in-situ recovery methods are used to highlight process and technology improvements that lead to enhanced overall performance. We demonstrate that the boiler, separator, and reservoir segments of the typical in-situ oil sands operation are key areas for improving the performance indicators of the production operations. Overall, solvent-based recovery is observed to be very promising from environmental and efficiency perspectives but very sensitive to the economics of the solvent being used. Following solvent-based methods is steam with non-condensable gas (NCG) addition. Steam and NCG injection offers a relatively balanced alternative with regards to the process performance indicators. Our results highlight the importance of integrating sub-surface and surface decisions when assessing innovations in oil sands process designs.

514279 Model-Guided Optimization of Industrial Azeotrope Distillation Column

Abstract:

A highly non-ideal mixture is separated using azeotrope distillation by introducing an entraining agent. The key components and process impurities in the feed along with the entrainer form multiple azeotropes.

The column exhibited chronic underperformance leading to significant loss of valuable chemicals. A rigorous model was developed using literature and proprietary data to represent the highly non-ideal coupled system. Although the model was not able to exactly match plant data, it was successfully used as a tool to improve the performance of the industrial column. In this presentation, the authors will be sharing the applied methodology.

513180 Knowledge Transfer and Support of New Technology Acquisitions

Abstract:

Successful acquisition of new chemical process technologies requires many steps both from a commercial and technical due diligence standpoint. Two of the most important steps are the meticulous transfer the entire knowledge base to the acquiring company followed by the development of plans for continued technology support after the commercial deal is completed.

In many instances, the technology seller provides limited technology support for a finite time after the commercial deal is complete, making it imperative that every effort be made to efficiently transfer as much of the technology knowhow as possible before support ends.

The plan going forward may include the retention of key technical employees that were intimately involved in the development of the technology and/or sales force employees that know the market and customer base, but ultimately the acquiring company must include the development of these critical resources within its own operating footprint.

Those resources include research and development, engineering, marketing and licensing as well as many other services key to the commercialization, sales and support of the newly acquired technology.

The entire process is illustrated in the cases studies of recent acquisitions by TechnipFMC.

513334 Optimal Selection and Design of a Methanol Process with Enhanced CO_2 Utilization

Abstract:

The recent discoveries of significant reserves of shale gas have spurred various monetization pathways including the production of methanol. There are several routes to convert shale/natural gas to methanol. The decisions for technology selection and process design were typically based on techno-economic criterial. Because of the growing interest in reducing the greenhouse gas emissions of gas processing, carbon footprint is emerging as a primary criterion in the selection and design of the process. The purpose of this work is to carry out process synthesis, selection, and design of a shale-gas to methanol process with multiple objectives including profitability and carbon footprint. Several reforming pathways along with the associated downstream processing are considered. In addition to the screening of individual types of reforming, the study also considers combined reforming of methane which incorporates three types of reforming, namely steam reforming, partial oxidation, and dry reforming. It is desired to consider existing technologies and to also modify the design of an existing methanol technology to include combined reforming. Utilization of CO_2 is addressed while maintaining the suitable syngas ratio for methanol synthesis, and minimizing CO_2 emissions, waste water generation, and energy requirements of the overall process. A methanol from natural gas production plant with a capacity of 5000 metric tonnes per day is considered as the base case scenario. The plant employs conventional steam reforming to generate syngas for methanol synthesis. Aspen HYSYS along with optimization software (LINGO) are used to simulate the process scenarios and to determine the optimum reforming configuration, respectively. Mass and energy integration are performed to improve the sustainability and profitability of the process, through identifying water and heat sinks in the design. A techno-economic analysis of the proposed design is conducted, including the cost of retrofitting, carbon taxes, and utility requirements.

AUTHOR INDEX

AUTHOR INDEX

AIChE
120 Wall Street, FL 23
New York, NY 10005-4020

CURRAN ASSOCIATES INC.
proceedings.com

ISBN 978-1-5108-6446-7
9 781510 864467